TREASURE HUNTER'S HANDBOOK

TREASURE HUNTER'S HANDBOOK

JOHN E. TRAISTER

FIRST EDITION

FIRST PRINTING— APRIL 1978

Printed in the United States
of America

Library of Congress Cataloging in Publication Data

Traister, John E.
Treasure hunter's handbook.

Bibliography: p.
Includes index.
1. Treasure-trove. I. Title.
G525.T66 622'.19 77-18993
ISBN 0-8306-7996-0
ISBN 0-8306-6996-5

Preface

No matter where you live—by sea or mountain, in the big city or open plains—there are places to hunt treasures. There are old towns, mines, battlefields, dumps, beaches, rivers, streams, and oceans. There are hundreds of places near your own home. Coins, historical relics, old firearms, antiques, minerals, jewelry, silver, and gold are yours for the finding, if you know how to look for them.

This book teaches you how to find all these treasures—and more. It shows you how to select the proper equipment to suit your treasure-hunting interests, how to locate treasure sites, how to repair and maintain your equipment, even how to build some of your own.

And once you find your treasure, you'll want to know how much it's worth. There's plenty of information on that too: you'll learn how to evaluate your finds, how to care for them, and how to sell them for top dollar.

Contents

Chapter 1
The Lure Of Treasure Hunting

Since the beginning of recorded history, few thoughts have so captivated the imagination of mankind as that of finding a fortune—King Solomon's mines, the treasure of the sunken continent of Atlantis, sunken treasure galleons, lost gold mines. The list is endless. People in search of these treasures, whether mythical or real, have numbered in the thousands. The lust for these treasures and the power they would confer have caused mankind to undertake superhuman feats and scale the pinnacles of both good and evil. Countless thousands have been exploited, murdered, and made fools for such treasures, even though nothing may have been found in many instances.

It is estimated that there is over $500,000,000,000 in lost treasure throughout the world today and each year thousands still venture to the four corners of the earth in search of these treasures. A few of these adventurers are professionals—ones who use thousands of dollars worth of equipment in their search, hoping to find enough to pay their expenses. However, the majority of these treasure hunters are people like you and me who find treasure hunting an exciting and rewarding hobby or even a fascinating part-time job. It's a hobby that allows us to get out of doors where we are provided with physical exercise and a chance to get away from the rat race.

Fig. 1-1. On a Civil War battlefield your finds may consist mainly of Minie balls like these—valued at about 25¢ each.

Treasure hunting is no longer just a game for a few subsidized adventurers who want to get rich(er) quick by discovering lost gold mines or a pirate's cache. It's now becoming a weekend adventure for thousands of American families—with participants ranging in age from two to 90 years. And it's surprising just how few come home empty-handed!

The types of finds and the methods used vary almost as much as the types of people who are in search for treasure. But regardless of where you search, there is some type of "treasure" everywhere—maybe in your own backyard. Imagine the thrill of taking your family camping for a weekend near a cool mountain stream and panning $300–$500 worth of gold in your spare time. With an electronic metal detector perhaps you'll locate a jar of old coins or a chest of silver ingots or a container with money or valuable jewelry inside.

Many people have done just this and many more will do so in the future. Most of them will not find anything as valuable as a pirate chest loaded with jeweled trinkets and pieces of eight. Rather, most of their finds will be relics of bygone days—now antiques. These finds are obviously not as glamorous as pirate gold, but they are still valuable to collectors. How valuable? This depends entirely on what is found. The find could be just a few Minie balls found on an old Civil War battlefield. The Minie balls may be worth only about 25¢. However, on the very same battlefield you could find an officer's belt buckle worth $300 or a sabre worth up to $150 or perhaps some Burnside cartridges worth about $5 each (Fig. 1-1).

One treasure hunter on vacation in Washington state found a rare gold coin valued at $300,000; others have found dozens of old coins worth from $100 to several hundred thousand (Fig. 1-2). And recently a stockbroker from New York City found one of this cen-

Fig. 1-2. Mr. and Mrs. Harry Bowens of Spokane, Washington, using Garrett Electronic TRs.

Fig. 1-3. Roy Lagal, well-known treasure hunter and author, scans a beach for lost coins and jewelry.

tury's most significant archaeological finds with a metal detector that cost less than $200. The find? A bronze statue of the Roman emperor Hadrian. The value? According to Dr. Cornelius Vermeule, curator of classical art at Boston's Museum of Fine Art, the discovery was the equivalent of digging up a Rembrandt and is worth three to four million dollars!

The reasons for these significant finds are simple. The earth has been the repository for vast amounts of wealth over the centuries. Many people who buried their treasures died without telling anyone where their cache was located. But with the development of new electronic metal locators and similar devices, much of this lost treasure is being located by weekend hobbyists (Fig. 1-3).

Outlaws' buried loot has always been high on the treasure hunter's list. Many misers, eccentrics, and recluses have buried their wealth, too. Many times, when these people die the location of their wealth is lost. Often a local contractor finds their wealth when the house is being torn down. Or, once in a while, a treasure hunter combs the grounds with a metal detector and finds the treasure and claims it for himself (Fig. 1-4).

One newspaper account told of a man who had appeared to be poverty stricken and had died of malnutrition. Yet over $250,000 in cash was found hidden in his home. There is obviously a lot of this type of treasure just waiting to be found—perhaps by the weekend treasure hunter.

There is no need to wander in the dark either. History books are filled with accounts of plundering by pirates, stagecoach robberies, and the like. Often their loot was buried not far from where the action took place—the closest cove, the nearest cave. Most of these treasures are still unfound today because knowing the approx-

Fig. 1-4. Treasure hunters using metal detectors to scan an old home site.

imate location of the action is not enough. Perhaps the treasure was buried in an area of 10 square miles, maybe more. But the modern treasure hunter can use certain techniques that may pinpoint the exact location of much of this buried treasure. Many of these modern techniques are described in this book.

But you don't have to look under the earth or in the water to find valuable treasure. There are many items sought by collectors that may be found right on the surface. Searching for these requires no equipment or special skill. You just need an observant eye and a knowledge of what you're looking for.

Many collectors, for example, pay good money for pieces of old barbed wire found around old farms and ranches. Some samples have brought as much as $1 per inch. These same farms and ranches are almost certain to have their own private dumps where bottles, tin cans, canning jars, and other items were disposed of. These same items today can be worth money to collectors—from $1 to $5 and up apiece—and there are thousands waiting to be found. All the weekend treasure hunter needs is an inexpensive book describing what bottles and jars are valuable. Then he is ready to spend many pleasant hours looking for the "junk" that can be converted into cold cash.

For those who enjoy hiking in the outdoors, there are woodland treasures that can bring both profit and pleasure to the entire family. Valuable plants are one form of treasure the outdoorsman can collect during the warm months. Drug companies use various plants for manufacturing medicines and are willing to pay good money for the plants they need. Many treasure hunters collect medicinal plants while camping (Fig. 1-5).

While there are over 100 medicinal plants that can be sold for cash, ginseng is the leader—worth over $65 per pound of dried roots. Goldenseal and Seneca-snakeroot, two other valuable medicinal plants, grow in the same areas as ginseng; that is, in high open woods, usually in rich soil on hillsides or bluffs affording natural drainage and shade (Fig. 1-6). Any of these plants, if found in reasonably good patches, can easily net the collector $20–$30 of extra income in only an hour or so of digging.

If you like skin diving then you'll find a lot of uses for your skills in searching for underwater treasures. For example, during the Revolutionary War, more than 500 ships were lost off the east coast

Fig. 1-5. A camper searching for valuable plants.

Fig. 1-6. Ginseng: the most valuable of all medicinal plants.

of the United States, and most of them still contain valuable relics and treasures. Off the west coast there's also plenty of treasure to be found. On the Great Lakes, over 15,000 vessels have gone to the bottom or broken up along the shorelines in the past 300 years. There are even maps available giving the location of several of these wrecks.

Rivers and inland lakes are also loaded with treasures that would interest almost any weekend treasure hunter. Outboard motors, fishing tackle, rings, coins, and jewelry can be found in nearly every body of water in this country. You can use an underwater metal detector, an airlift device (similar to a vacuum cleaner), or even a magnet to recover most of these objects.

Perhaps you'd like to try your luck with a gold pan like the ones used by the forty-niners. Many families are doing just that. With a gold-panning kit—costing less than $10—they're panning nearby streams for gold dust during their spare time. Surprisingly enough, many of them are returning home with $50–$500 worth of gold dust—collected in a single weekend.

Others may be content to search for valuables in their backyards on a public beach after a hot summer's day or in football stadiums after a game or in parks or on carnival grounds or anywhere people gather. A reservoir near one treasure hunter's home yielded nearly $50 in silver coins, several wristwatches, dozens of rings, and over 1000 pennies. Two women reportedly found a chest in their backyard containing approximately $700 in cash using a newly acquired metal detector.

The services of treasure hunters are also sought on a professional basis. Land surveyors often ask people with metal detectors to locate hidden land markers. Insurance companies, facing claims on lost items, sometimes require the services of treasure hunters, too. Even law enforcement agencies have sought the services of treasure hunters to recover bullets shot into the ground. The list grows each month.

Though it is certainly true that many treasure hunters take to the hunting grounds just for the sake of getting outdoors, the greatest reason for this growing sport is for the excitement: the thrill of seeing a glittering substance in your gold pan, the anticipation of recovering a hidden object underground when your metal detector indicates a find, the enjoyment of flipping through a guidebook to identify a patch of wild plants you have found. Maybe it's the same feeling you get when you pull the handle on a slot machine. There is always a chance of hitting the jackpot.

Chapter 2
Looking For Land Treasures

I would venture to say that there is probably at least one large treasure cache—still unfound—within a 50-mile radius of every community in the United States. No doubt there is some kind of treasure to be found *in* and *around* every community in this country. In fact, you can probably find at least something of value around any old house.

These land treasures will vary in size and value from a can of old coins found at the bottom of an old cedar fence post to a chest of gold bars hidden by 18th century bandits. In between this spread lie so many valuable items that all treasure hunters are constantly guessing about what they will dig up next. Antiques, rings, knives, firearms, old tools, and old coins are just a few of the exciting finds commonly uncovered with a metal detector.

Many of the old discarded bottles which were common in bygone days are now worth money to collectors—a lot of money if the right ones are found. It is entirely probable that one old farm dump could contain hundreds of old bottles worth from $2 to $5 apiece; some very rare ones have sold for up to $2000 and more. So the next time you notice some people poking around and digging into some trash dump, you'll know what they're looking for.

These same dumps have also contained old tools that are now priceless; even old firearms have been recovered. Of course, they are not operable, but as collector's weapons they still have great value. Even old telephone insulators can sometimes be found in old dumps. They can be worth up to $500 each.

Recently, a lady was rummaging around in the attic of an abandoned home just before it was to be demolished and located a cigar box full of stamps which dated back into the last century. The total value of these stamps had not been determined at this writing, but a rare United States one-cent stamp recently brought $260,000.

A doctor in Austin, Texas, discovered an old treasure map during his research and immediately set out to find the cache. Surprisingly enough, he located the treasure on Padre Island, Texas, in the form of two chests full of coins valued at more than $75,000. Another man exploring a Delaware beach after a storm picked up a goatskin bag containing over $50,000 in 17th century coins. In a plowed field in Arizona, two boys found the $85,000 loot from a robbery. The list of land finds is endless.

GETTING STARTED

Let's assume that you have just purchased your first metal detector. It can be a build-it-yourself model for $10 or so (see *How To Build Metal/Treasure Locators*, John and Robert Traister, TAB No. 909), or one of the better models on the market costing $400 or more. First, read the manufacturer's instructions contained in this book. This is extremely important because familiarity with the instrument can make the difference between a find and a miss.

Begin by taking the detector out of doors and testing it on coins, rings, etc., until you become familiar with the sound emitted for each item. If you listen carefully (this will take time to learn), you will begin to tell the difference between a large and small object by the sound made by your detector. You may even be able to tell the difference between worthless junk like wire and nails and valuables like coins and rings.

When you have practiced for an hour or so with the detector, try it out on some real game. Begin at your own house, searching along walkways, especially where people get in and out of cars. Search these areas carefully. Chances are you're going to find a lot of pull tabs, chewing gum wrappers, and the like, but you should also find a coin or two. You may even find a ring and a few car keys. Chances are none of these items will be worth much, but you'll be getting experience and having a little fun besides.

Fig. 2-1. In addition to finding large quantities of coins and jewelry, the coin hunter may uncover many old tokens such as these.

When you have combed your own premises carefully, and found (and set aside) every nail, pull tab, coin, ring, and any other metallic objects, you can search other areas. You can comb places where your chances of success are greater. The reason it is not recommended that you start in these places is that you would probably miss a good many of these finds if you were not familiar with your detector. Some other person may see you and get the same idea. But with a little practice in your own yard beforehand, your chances of finding more will be greatly improved.

You will find the most coins around areas where large crowds have gathered—school grounds, carnival grounds, sports arenas, etc. In addition to finding large quantities of coins and jewelry, you will also probably find many tokens like the ones shown in Fig. 2-1. These are typical finds by today's coin hunters.

Perhaps you'll be content just to hunt coins in your spare time. This type of treasure hunting can be rewarding in itself, because the average "coin shooter" finds around 5000 coins a year in his spare time. In *face* value, the total amount of these coins seldom runs over $500. But let's consider the *collecting* value of these same coins. One 1872 Indian cent found around an abandoned church in northern Virginia earned the finder $20 from a coin collector. A 1914 D Lincoln cent found nearby brought around $35 from the same buyer. Even many recent coins are worth from $25 to $100 each! All of this means that a dedicated coin shooter can possibly earn an average of about $3 for every coin he finds. It doesn't take a mathematical genius to figure that finding 5000 coins a year can mean $15,000 annually. Of course, every coin shooter is not going to earn this much, but the possibility is still there.

You may even accidentally bump into some other valuable object while searching for coins. Take, for example, the two children who lived near Vicksburg, Mississippi. They happened across a Civil War treasure that was appraised at $165,000.

HISTORICAL RELICS

After you have cut your teeth on some coin shooting around one of the parks or school grounds near your home, you'll probably want to begin looking for other valuables. A good place to start looking will be around old home sites. Such places are relatively plentiful, especially in rural areas. Your finds may include items like those pictured in Fig. 2-2—old firearms, barbed wire valued at over $1 per foot, Civil War relics, and the like.

If you want to explore the wars of the past, there is hardly an occupied corner of this country that has not seen armies in battle. Most of the well-known battlefields have been combed over and over

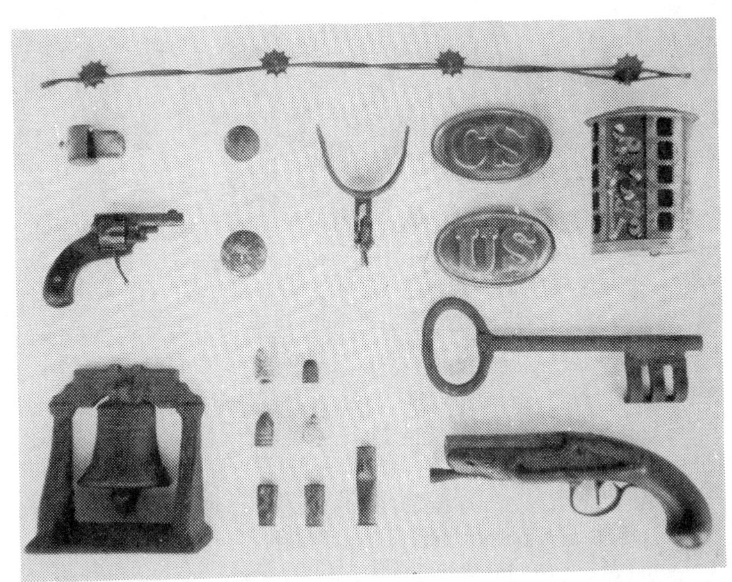

Fig. 2-2. These are valuable relics found around old home sites with a metal detector.

Fig. 2-3. Wayne Austin of Shenandoah, Virginia, searching for relics on the Milford, Virginia, battleground.

again, and it is assumed that most of the relics that were in these areas are now depleted; or else, laws prohibit any treasure hunting. Still, there are dozens of lesser known areas where all kinds of war relics can be found. It just takes some research into records before you start digging in the field.

The most famous battlefields are state or national parks and are fully protected by law against treasure hunting or digging of any sort. But what about the less famous ones—those that are now privately owned?

Wayne Austin of Shenandoah, Virgina, (Fig. 2-3) did some research at his local library and uncovered some clues on battles near his home. Most of these battlefields were on private property. So all Wayne had to do was to obtain permission—often offering the land owners 50% of the find—to hunt for the relics. With a government surplus metal detector, he was able to uncover dozens of lead bullets—both dropped and spent—in only a couple of hours. A later visit turned up a spur, a belt buckle, an ammunition box cover, and other Civil War relics. One of his finds is shown in Fig. 2-4.

By doing some research of your own, you may be able to turn up a battleground near your home that is still untouched by relic hunters. Your finds could include an eight-inch Parrott shell valued between $100 and $125, an oval belt plate worth $1500, a glass inkwell valued at $40, lead bullets worth up to $25, a coat button which can bring up to $300 from collectors. All of these are typical finds on any Civil War battlefield.

Detailed records of wars and battles can be found in most libraries. Such records can tell you exactly where the fighting took place. If you need some help in locating these areas, talk to your local treasure-hunting supplies dealer. He will either tell you where the

Fig. 2-4. A clump of Minie balls and pistol bullets found on a Civil War battlefield.

hunting areas are or put you in contact with individuals who will know.

The University of Alabama in Birmingham even offers a college course in treasure hunting and, by the time this book is published, other colleges in other parts of the country may be doing the same. The course at Alabama is slanted toward relic hunting and archaeological exploration, but it also includes the use of metal detectors.

The Civil War era is also noted for its looting by both sides. The loot included family treasures, jewelry, firearms, money, gold, silver, etc. Much of this treasure was hidden in caves. Those who survived the war often returned to collect their cache later; those who didn't survive left the treasure to be found by others. This is where you come in.

A trip to the local newspaper office to look over back issues of the paper may turn up some clues about the location of caves. Some friendly persuading of the older residents in an area will certainly uncover many clues to the location of caves as well as battlegrounds, forts, Indian villages, military routes, and other probable treasure-holding areas.

Most of these caves will have very small openings and can be the size of your bathroom or as large as a football stadium. The smaller sizes will be the most common. If you choose to search these caves, be sure to go with a buddy and use all precautions against accidents.

More than likely none of the treasure—if any—is going to be in the open passages of caves. Rather, it will be buried under a dirt floor, in the walls, or perhaps under a pile of rocks. This is where your metal detector comes in. Comb every inch of the cave with your detector (including the ceiling) and you might come up with a treasure worth enough to support you the rest of your life.

Don't overlook the abandoned homes near battlefields (Fig. 2-5). Again your metal detector will come in handy for locating hidden relics or family fortunes. A Gunnison .44 caliber Confederate revolver, for example, was located under the floor of an abandoned house in North Carolina in August, 1968, with the aid of an electronic metal detector. Its value was estimated at over $700. Another treasure hunter found a tin coffee can containing several hundred dollars worth of paper money and coins. The money was worth much more than its face value.

Fig. 2-5. Old abandoned homes are a good place to look for hidden relics or family fortunes.

GEM AND MINERALS

No survey of the treasures found on land would be complete without the mention of gems and minerals—probably the first treasures sought by mankind. Before you begin, however, you should acquire at least a basic knowledge of what to look for. The bibliography at the end of this book can help.

Besides gold, which is found in practically every state of the union, there are all kinds of rocks, gems, and other minerals found virtually everywhere. You don't have to go to South America to find emeralds, or to South Africa to find diamonds or to India to find rubies. There are still many places in the United States where gemstones can be found—even lying on the ground. You will have to know what to look for and be willing to devote a lot of time to searching.

TURNING PRO

Maybe you will become so interested in the hobby of land treasure hunting that you'll turn pro by starting a business of your own. In fact, anyone with a metal detector and a good knowledge of its use can start a lost and found business on a part-time basis. Such a part-time business may eventually turn into a full-time venture. Every day someone loses a valuable ring or other piece of jewelry, many of which are priceless, and the owners usually offer a sizeable reward for the return of these items. In cases like these, the reward will be your treasure.

To get started in a service like this run a small ad in the classified section of your local newspaper. Read the lost and found columns; tell insurance agencies of your service. Once you get started, other possibilities will turn up. One treasure hunter was having gas put in his car at a local service station when the owner of the station noticed a metal detector in the back of the car. It seems the owner had lost the front sight off of an obsolete handgun while target practicing behind his station. He asked the guy with the metal detector if he would take a few minutes to help him find the sight. The sight was located in a matter of minutes. The guy with the detector got a free tank of gas.

You will probably get more business if you operate on the no-find no-pay basis. This way the customer knows he will not have to pay for anything. If you find the item, a fee amounting to 50% of the item's wholesale value is not unreasonable.

When you take on such a job, make certain you get as much information about the lost object as possible. Find out the boundaries of the search area. If possible have the owner of the lost object go with you to point out the boundaries. Good leads and a good knowledge of your metal detector will make collecting the reward almost a formality.

Chapter 3
Equipment For Treasure Hunting On Land

The equipment needed for hunting treasure on land may be as simple as a pair of gloves for rummaging around dumps in search of old bottles. Or it can be as elaborate as a dredging machine costing over $100,000. The important thing is that you have the right equipment for the job.

METAL DETECTORS

Metal detectors range in price from $25 to over $400. The sensitivity of the instrumentation usually increases with the price, but for those who plan to hunt regularly those detectors ranging in price from $150 and up will be the best bet. They will quickly pay for themselves in more consistent finds.

Selecting a Metal Detector

There are three types of metal detectors currently available: the BFO (beat frequency oscillator) detector, TR (transmitter-receiver) detector, and the VLF (very low frequency) detector. For many years the BFO was the best type of metal detector available, but recent developments have made the TR instruments—sometimes called IB or induction balance detectors—much more sensitive, easier to use, and more reliable in most cases.

The BFO detector (Fig. 3-1) should be the choice of those who plan to engage in a wide variety of treasure hunting—especially in

Fig. 3-1. This BFO metal detector is manufactured by ECE, Inc., of Belen, New Mexico. It operates on a frequency of 10 kHz, resulting in good performance with excellent ground penetration and discrimination.

search of minerals or for use over rough terrain. Many models are available with a discrimination feature which allows you to "tune out" worthless items like tin foil, nails, pull-tabs, etc. The chief disadvantage of this type of detector is with the smaller items like rings and coins; it is a much more difficult instrument to learn to use than the TR or VLF. Inexperienced operators fail to recognize the weaker signals of small objects and therefore cannot achieve the depth they could with the TR or VLF instruments.

For those who plan to engage primarily in coin hunting, with occasional hunts for relics and buried caches, the TR detector is the one to choose (Fig. 3-2). This type of instrument is easy for the beginner to learn to use. It has a quick response, a near silent tone (except when a find is indicated), and discriminating abilities. It is hard to beat for coin hunting.

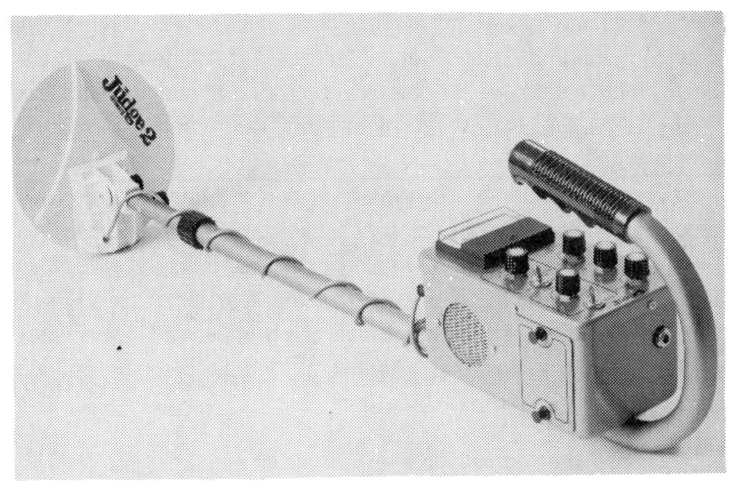

Fig. 3-2. A TR metal detector manufactured by Compass Electronics Corp., Forest Grove, Oregon.

However, one of the TR's disadvantages is that when it's used over rough terrain that contains minerals, the instrument is bombarded with false signals which mask true signals caused by buried metal objects. You can correct for this somewhat by tuning the detector down so that it does not react so strongly to the mineralization. Unfortunately, this also means passing up many finds that may be valuable.

The most recent metal detector development is the VLF (Fig. 3-3). This instrument is, without a doubt, the deepest seeking detector available. It is also the easiest to use and the most expen-

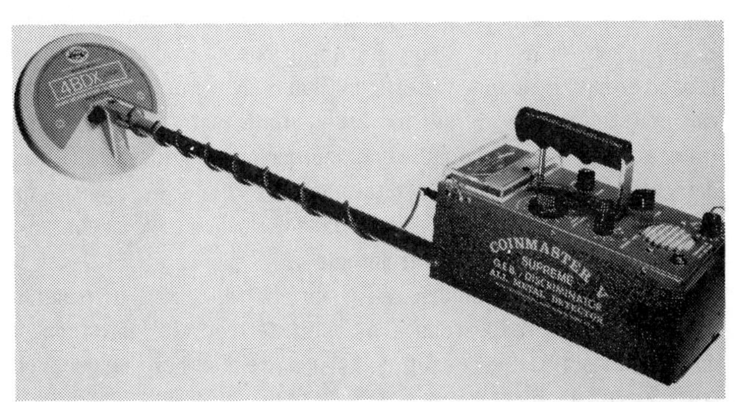

Fig. 3-3. The most recent metal detector development is the VLF type. It penetrates deeper and is easier to operate over mineralized ground than any other type of detector. This one is manufactured by White's Electronics, Inc.

sive. As sensitivity goes up, so does price! This type of detector will detect anything that will conduct electricity. It will detect things at depths which other types of metal detectors cannot approach. This type of detector also excels—with a few restrictions—in mineral prospecting.

The VLF detector, however, is not recommended for use around areas where a lot of metallic junk is present. This is because the VLF has a wide range and thus cannot discriminate between all the objects it detects. The weight of the search coil on VLFs is usually much greater than that of coils on other detectors, and, after a day's hunt, you know that you have carried the detector around!

Theory of Operation

Though the controls on every brand of metal detector will be slightly different, the basic operating principles of each type (BFO, TR, or VLF) are essentially the same.

Metal detectors indicate a find by a change in the audible pitch they emit. In general, a signal is broadcast down into the ground and is then received by the detecting part of the instrument. If this signal is not altered by the presence of a metallic object, the continuous tone from the speaker will not change. The presence of metal, however, will alter the signal and register as an increase or decrease in pitch from the speaker.

Depth of Detection

Depth of detection depends upon a great number of factors. A large object can be detected at greater depths than a smaller object. If a metal object has been buried in the ground for a long time, the ground around it becomes altered by that object. For instance, a tin can that lies in the ground for a long time rusts; this oxidation radiates out and away from the can, increasing the area in which the detector can register a find. To the detector, the object appears to be larger than it actually is. This is called the halo effect. Even coins buried a long time increase in detectability.

Minerals in the soil can cause surface reflections or mineral reflections that greatly reduce or even destroy the instrument's ability to detect. Damp or saturated soil may either increase depth of detection or decrease it.

Even the attitude of an object can affect depth of detection. A coin buried on edge is more difficult to detect than one lying flat. This

Table 3-1. Results of a Laboratory Depth of Detection Test.

Test Objects	Distance Detected from Search Coil
Mercury Dime	6 1/2 inches
Copper Penny	7 inches
Silver Quarter	7 1/2 inches
Gold Wedding Band	8 inches
Silver Half Dollar	9 inches
Silver Dollar	11 1/2 inches
3 1/2 inch Jar Lid	17 inches
Old Adz Head	18 1/2 inches

is because the greater an object's surface area exposed to the detector field, the easier it is to detect the object at greater depths.

The shape of an object also has an effect on depth of detection. Rings, for instance, can be detected at greater depths than a coin of comparable size. This is the reason that pull tab rings cause so many problems with nondiscriminating instruments.

Table 3-1 will give you an idea of how depth of detection can vary according to the shape and size of an object. The table gives the results of a laboratory depth of detection test on one brand of metal detector.

The TR Detector

A block diagram of a transmitter-receiver metal detector is shown in Fig. 3-4. Within the electronic housing there is a transistorized transmitter and receiver circuit coupled to a transmitter and receiver antenna network which is housed in the search head. When the detector is energized an electromagnetic field is developed around the transmitting coil (Fig. 3-5). The receiver coil is inductively balanced to the transmitter coil and the receiver reception is adjusted by controls on the instrument. When a ferrous or nonferrous metallic object is introduced into the electromagnetic field an unbalanced state will result between transmitter and receiver coils (hereafter referred to as the "loop"). The receiver section interprets the unbalanced condition and it is heard as a change in pitch in the speaker or seen as a deflection on the meter.

The simplest form of detection involves observing amplitude changes in the receiver coil. Going one step further and examining

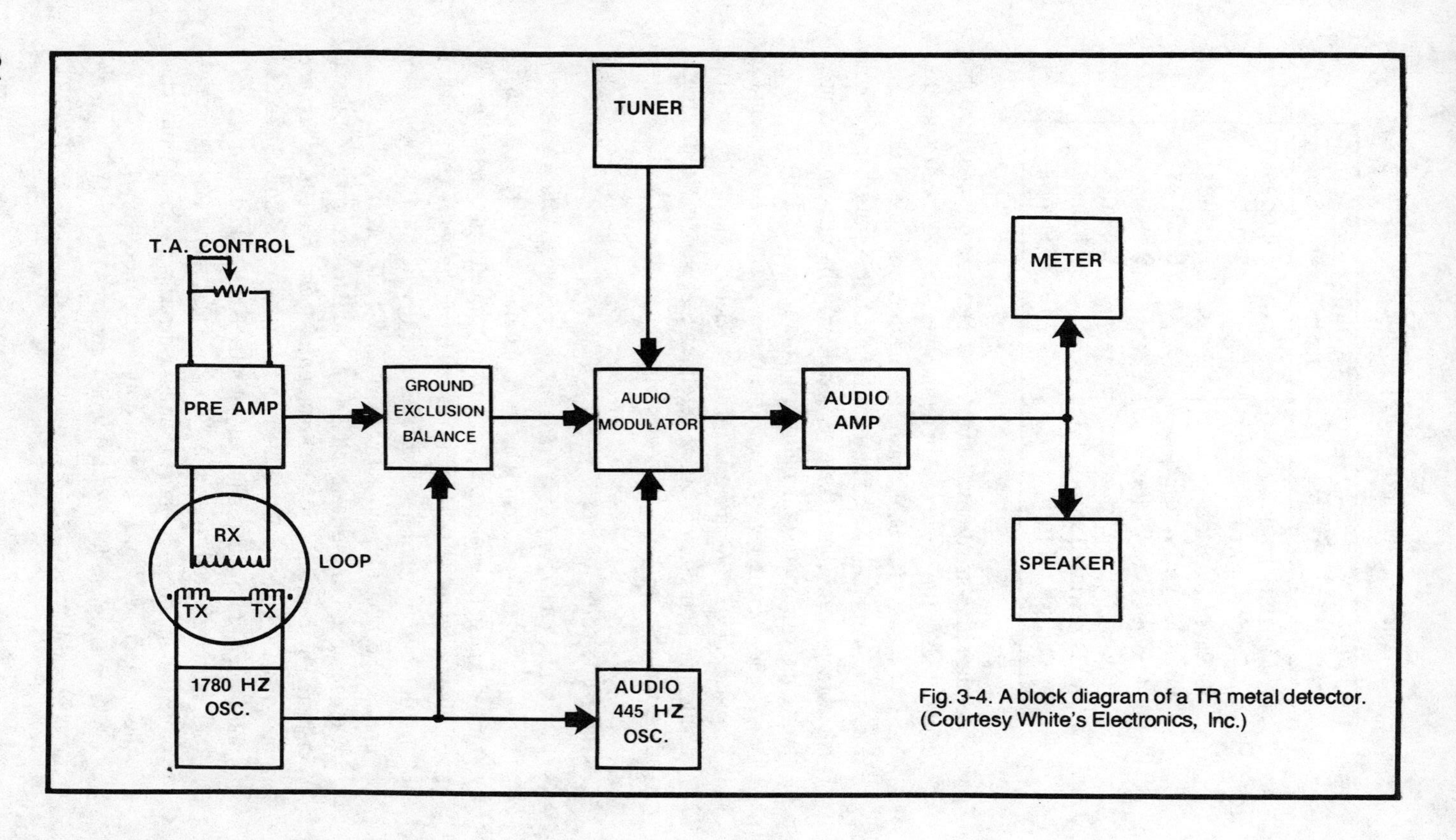

Fig. 3-4. A block diagram of a TR metal detector. (Courtesy White's Electronics, Inc.)

the phase change in the receive signal, we can begin to distinguish various classes of metal objects. Most bottle caps belong to the ferrous classification and are distinguishable from pull tabs which are in the thin plate aluminum classification. Both bottle caps and pull tabs can be differentiated from highly conductive thicker objects such as coins. The discrimination control sets up a circuit condition so that the highly conductive thicker objects (coins) cause an increase in the receive signal level and all other objects cause a decrease. Since coins cause less phase change than both ferrous and thin, less conductive materials, they can be distinguished as a class by themselves.

The ground control of the TR is primarily a tone control so in highly mineralized ground the background static can be decreased. The ground control does not appreciably decrease the instrument's sensitivity to coins. In highly mineralized ground decreasing the ground control to normal will actually increase the depth you can search for coins because it decreases the masking static sound of the minerals in the ground.

Whether a detector responds to minerals or metals is primarily a function of the object size. There are actually two electrical processes involved—induction and surface currents—which can occur in both metal and minerals. The induction process dominates in

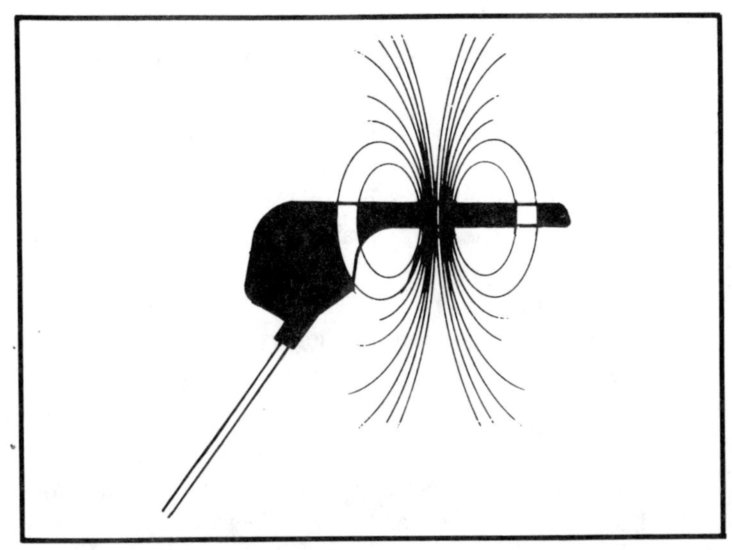

Fig. 3-5. A side view of a search head with electromagnetic field.

Table 3-2. Approximate Depth Ranges of a TR Detector With 8-Inch and 12-Inch Loops.

Object	8 inch Loop	12 inch Loop
Penny	6 – 8 inches	6 – 8 inches
Silver Half Dollar	8 – 11 inches	10 – 12 inches
Silver Dollar	10 – 12 inches	12 – 14 inches
Pistol	15 – 17 inches	18 – 21 inches
Bottle Dump	48 inches	60 inches

metals and surface currents dominate in minerals. However, the smaller an object, the more predominant the surface currents in relation to the induction process. Thus, any metal object that is 1/4 inch or less in diameter will register as a mineral.

The TR metal detector will detect most ferrous and all nonferrous metals. The electromagnetic field will penetrate through wood, rock, adobe, ice, snow, soil, nonreinforced concrete and asphalt and a lot of other materials. However, concrete is difficult to work over because it is highly mineralized, and the TR will not detect glass, gems or sulfides. An object must have a conductive property in order for it to affect the loop's magnetic field.

Table 3-2 shows the approximate depths at which a TR can detect various objects using 8-inch and 12-inch loops.

Figures in Table 3-2 are approximate and are conservative for most soils. Coins have been reported being found at depths of 18 inches using an 8-inch loop; however, these finds were made under ideal circumstances.

Mineralization can have a very pronounced effect on the field generated by a detector loop. Unfortunately, no two areas are alike as far as mineral content in the coil is concerned. When I refer to mineralization I am speaking of salts, iron and magnetic mineral particles found in most soils. In the western states and a few southeastern states mineralization can be high and may greatly reduce depth of detection. In many other states mineralization is low and greater depths can be attained.

Using Your Detector

After reading the manufacturers instructions on tuning your metal detector properly you are ready to begin using it. Hold the detector with your arm slightly extended in front of your body as shown in Fig. 3-6. Don't bend or stoop but stand in an upright position that is comfortable. By holding the detector out slightly from your body you can observe the slightest difference in response. Adjust the telescoping coil rod out or in until the search head is about 1/2 inch above the ground. Adjust the search head so it is *parallel* to the ground.

With the detector in the correct search position, swing the search coil gently side to side (Fig. 3-7). Slightly overlap each sweep as you move forward or backward. Make sure the search coil stays approximately the same height (1/2 inch) above the ground and don't allow it to lift at the end of the sweep. Lifting will cause false readings. Do each sweep slowly, about three seconds from one edge to the other.

When your detector signal is the strongest and the search coil is directly centered over the find you have pinpointed the target. When the detector signals a find slowly move the coil forward and backward, then side to side, until the signal is the most intense. Your find will then be directly beneath the center of your search coil (Fig. 3-8).

To make sure you search an area thoroughly, without gaps, you may wish to lay out the search area in a grid pattern. Stake out two parallel strings about three feet apart (the length of your sweep) and sweep this area thoroughly. Next, move one string over and make a new search area. Do this until the entire area has been covered. You may also grid an area mentally using trees, rocks, buildings, etc., as points of reference.

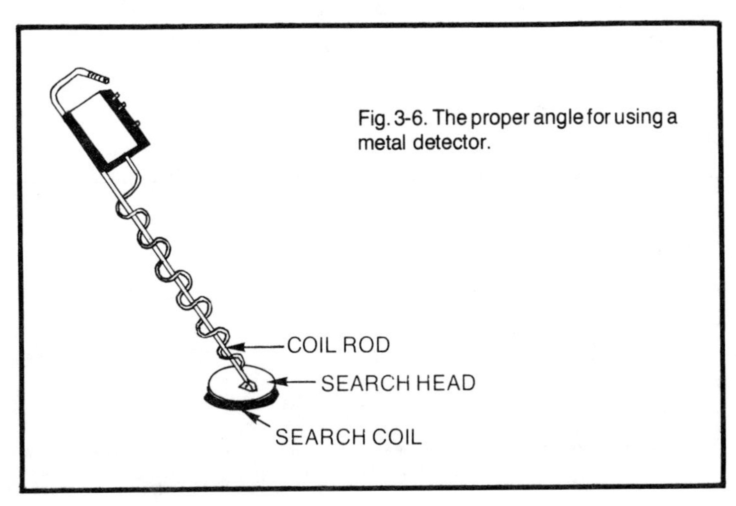

Fig. 3-6. The proper angle for using a metal detector.

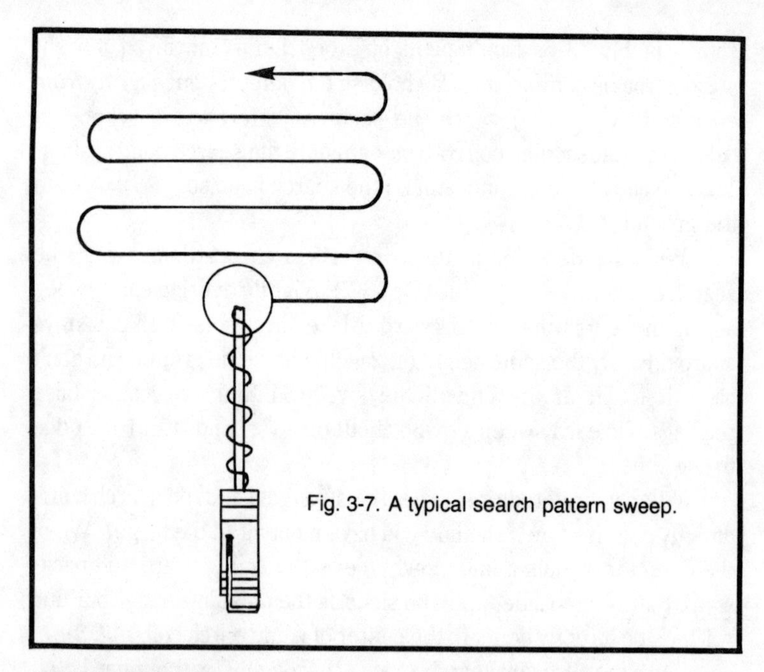

Fig. 3-7. A typical search pattern sweep.

OTHER EQUIPMENT

The equipment you take along on your treasure-hunt outing will depend on the treasure you're looking for, the terrain over which you will be searching, and how long you plan to search.

A backpack is useful in any kind of land treasure hunting. It serves as a container to carry equipment and any finds that may turn up. The backpack shown in Fig. 3-9 is an 18-inch pack basket. Such a pack sells for less than $20 and is not only large enough to carry all the required gear, but it is also substantial enough to double as a relatively comfortable seat or a support for a camera. Due to its shell-like construction this type of pack also offers excellent protec-

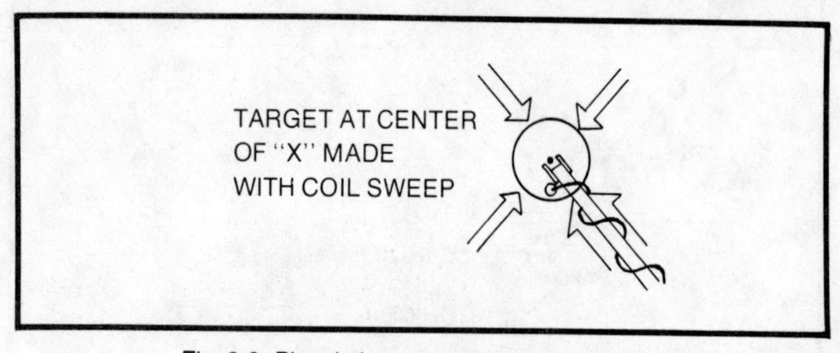

Fig. 3-8. Pinpointing a target with the search head.

Fig. 3-9. A pack basket like this one is ideal for carrying equipment on treasure hunts.

tion for all items within the pack; rawhide laces can be threaded through the basket weave to secure larger items to the outside of the basket.

One disadvantage of this type of pack, however, is its lack of individual compartments. Small items such as boxes of matches, film, and the like tend to slide down between the larger items regardless of how carefully they are packed. This means that if you need any of the smaller items while in the field, many of the larger items will have to be dumped first. However, a small canvas bag with pockets can hold smaller items. This bag can be placed on top of everything else in the pack and retrieved instantly. A surplus GI cartridge belt with pockets and flaps may be perfect for this.

Since much treasure hunting is done in warm weather in remote areas where rattlesnakes and copperheads may live, a snakebite kit should be one of the first considerations. You can buy a small snakebite kit that is smaller than a pack of cigarettes. It contains a constriction band, a razor blade, a suction device, an adapter tip for small areas like between toes, ammonia, and Band-Aids. Recently, however, doctors have been recommending a freeze method called Snakebite Freeze. This kit contains two packs of instant chemical freeze, two latex constriction bands, and one insulated Neoprene wrap with Velcro fasteners. Both kits are popular. But you and your doctor must decide which one suits your needs best.

A lightweight plastic poncho which folds up to about a 6-inch square is another handy item. It will keep you dry during a storm, ward off chilly winds, convert into a makeshift lean-to, and protect you from the damp ground during a rest period.

Dry matches are good to have on any outing. Waterproof matches can be purchased for less than $1 per carton and a box should be kept in the pack or your pocket at all times. If you can't find any waterproof matches at your local camping or sporting goods store, it is easy to make your own. Merely melt some paraffin over a fire, let it cool a little, and then immerse regular wooden matches in the paraffin for a moment. Remove the matches from the paraffin. Let them cool. The paraffin layer around the matches will keep them perfectly dry and does not affect their lighting capability. Another method is to dip the matches in shellac and let them dry.

Usually a compass is unnecessary if you are familar with the terrain in which you are hunting. However, in unknown country a compass is a must.

You may also want to take along a little spiral-bound notebook and a pencil or pen to make notes of your finds or the terrain.

You can avoid many miserable hours while in insect country by buying and using a good insect repellent. A small can will easily fit into your pack. In areas where insects are extremely annoying a head net is also advisable.

A flashlight is a lot of added weight to your pack and many treasure hunters do not bother to carry one. However, at times a small flashlight can be a big help. Dark crevices can house a rattlesnake. A quick search with a flashlight may help you avoid an accident. And a good light is priceless if you're making your way back home after dark.

If you know you're going to be hunting in areas where there is plenty of clean drinking water from streams or surface springs, a canteen or thermos is a lot of added weight and can be left behind. But if you're going to be in areas where the quality of drinking water is doubtful, a canteen or thermos can come in handy.

A good digging tool is a must. The type will depend on what sort of items you are looking for. For relic hunting on Civil War battlefields, a small screwdriver, a hunting knife, or a lightweight digging tool is all you'll need. Of course, if you are looking for large caches then a pick and shovel are a must.

SPECIALTY TOOLS

Special tools are needed to hunt for gems and minerals. These include a lightweight, short-handled pick, a metal probe, a pocket magnifying glass, a rock hammer and a pair of goggles. Another useful item is a "rock-scoop" which is a canelike stick with a scoop on one end. It's used for picking up individual rocks without bending over. It also allows you to poke into holes and crevices without risking snakebite.

Chapter 4
Land Treasure – Hunting Sites

If you go outside your house right now and swing a metal detector over your lawn (after first becoming familiar with the detector's use), chances are you would locate something of value—a coin dropped by one of the construction workers who helped build your house, a class ring, maybe an antique metal button. But you can sweep over your whole yard in no time. You'll then have to look elsewhere if you want to continue treasure hunting.

Although millions of coins, rings, relics and artifacts have been found by treasure hunters in this country over the past decade, the good hunting spots are yet to be discovered. Many people think that most of the better spots have been "searched out," but this is not the case. There is far more treasure *still* in the ground than has been recovered in the past 100 years. And often you don't have to travel hundreds of miles to find it either.

FINDING HUNTING AREAS

The best areas for locating coins, rings, jewelry, and the like are places where there have been large gatherings of people. For example, carnival sites, recreational parks, city parks, swimming holes, playgrounds, race tracks, school grounds and many, many more. These areas will offer many targets for your metal detector, and you will find one or more of these places within a short distance of your home.

An old field on the outskirts of your town may now be a pasture, but it could have once been a regular meeting place for tent shows or carnivals years ago. Check with the older citizens of your community and find out where these gatherings took place.

Obtain books from a nearby library about the history of your area. The treasure hunters who have the greatest success at finding treasures are those who take the time to do some research before they begin hunting. For example, a treasure hunter recently purchased some land in northern Virginia near a small village named Overall. He obtained several history books about the Civil War battles in the area, hoping to locate some battle sites on which to search for Civil War relics. While looking over some maps drawn by the Corps of Engineers during the period, he recognized the names of some nearby towns. He studied the maps further and finally realized that his village (Overall) was once called Milford, Virginia, and that a large battle was fought there. One map showed the location of trenches, batteries, and skirmishes. By comparing the map with a modern topographic map the treasure hunter discovered that his home was built right on the battleground. He immediately acquired a metal detector and during a few weeks of searching he had uncovered about 2000 Minie balls, an officer's spur, several artillery shells, and a rifle barrel—right in his own backyard.

So read all the historical information you can find about your area or about the area you intend to hunt on. Talk with people; gather all the clues you can.

Old abandoned home sites are excellent places to find valuables of all types. If you can't find the location of such places from old county maps, ask around. There are many old abandoned home sites all over the country and it is a hard matter, indeed, to search around one of these sites and come away empty handed.

In the old days, many people didn't believe in banks and most had some safe place to hide their money and other valuables. Many of these people died without ever disclosing the location of their cache. Much of this treasure still remains unfound. A person with a metal detector and a little patience could uncover a lot of these buried treasures if the location of the home sites were known.

What about gold, silver, and other minerals? Where do you look for these? Streams are a good place to hunt, but not just any stream will do. You must find streams that are *suspected* of containing valuable minerals. To find out which streams are which, you might have to look through back issues at a local newspaper. Or you may have to check with a geologist who knows the area.

If you will do approximately one week of research about the history of your area (if you live in the country or a small town) or an area that you plan to hunt, you probably will discover enough places to treasure hunt to last you for months—if not years.

Take a notebook and visit one of the older newspaper offices near the area. Most offices will have back issues of every newspaper they have published. You won't be able to take them out, but you can look through them in the office. Make notes of all fires involving homes that were completely destroyed, homes that were sold "on the block" for tax payments and any other items that would indicate that a home or a town was abandoned. Chances are these sites are still around and some valuables are still around also. The land on which these homes were built may be used for grazing land. Locate these places, ask permission from the land owners, and search them out good. The land owner himself will probably be able to show you the exact spot where the old structure stood.

Next, pay a visit to your local library and glance through every book you can find on local history. Make a note of any old maps in these books as well as the description of any homes that were destroyed. For example, one history book states that a cavalry detachment during the Civil War followed the Shenandoah River south from Front Royal, Virginia, stopped off at a village named Bentonville and burned the homes of several local farmers when they refused to give the troops food. This statement, in itself, would be of little help in pinpointing the *exact* location of the homes in question, but further research, either in the library or in the area, could lead you to the very spot.

Say you wanted to locate the Bentonville ruins. You could start at Bentonville (as the record states) and walk north towards Front Royal, looking for any signs of old foundations, etc. You could even talk to several farmers along the way; many of them are probably ancestors of those whose homes were burned.

History books can reveal the location of old Indian villages, abandoned mines, mills, factories, battlegrounds, and dozens of other places where the treasure hunter can locate valuable finds. If you research carefully, you may discover the location of treasures that have been left undisturbed for years.

For example, one Civil War buff recently discovered an early Civil War history book in a university library. The text was studied religiously until he came upon a description of a battle won by the Union Army. After the fighting the Confederate prisoners were required to turn over their assortment of weapons. The newer rifles were transported behind the lines, probably to be used by the Union Army if they were needed. The remaining 20,000 older rifles were used to corduroy a half mile of muddy road so the Union's cannon and wagons could be pulled over it without bogging down. Obviously, these rifles sank into the mud and were soon forgotten. Since the book gave the exact spot of the battle, it didn't take long for the treasure hunter to recover the rifles. The rifles were valued at fifty to several hundred dollars each—depending upon their condition.

You can obtain publications giving the basic facts on nearly any type of treasure you care to mention; maps are even included, but the exact spot of the treasure could be anywhere within a 1000-acre area. A list of books giving clues regarding the whereabouts of treasure sites is included in an appendix in back of this book.

SOME DO'S AND DON'TS

Sometimes *finding* a hunting area is not as important as being able to hunt on it day after day. It takes days to hunt some treasures. Following a few simple guidelines can insure that you keep on good terms with property owners so you can return to finish the job.

When searching on private property, always obtain permission from the property owner before you begin your search. If you are tracing down, or expect to locate, a large or valuable find on the property, it is wise to make the individual property owner aware of your desires *before* you begin the search. Sometimes, by properly identifying yourself, you may be able to gain permission to hunt where others have failed. The property owner may even have some ideas that could assist you in locating more valuable finds. Remember, he may be more willing to let you hunt on his property if you offer him part of the find.

Always respect the property. Never leave any hole or surface mar on the property. Clean up as you go along.

If you have permission to sweep your metal detector over a lawn you should use a screwdriver not more than six or eight inches long and 1/4 inch in diameter. A larger instrument could damage the lawn. You will probably be looking only for smaller items like coins and rings anyway, so larger tools are unnecessary. Never probe over six or eight inches deep for an object in areas like these and do not use knives or big probes that could tear up the lawn and ruin the possibility of future treasure hunting.

Always fill in holes you have dug in probing or digging for a find. Holes are unsightly as well as dangerous for passersby.

The method of removing an item from the soil is dictated by size of the item itself. As an example, let's assume you're searching for an old coin. First, use a screwdriver or similar tool to pinpoint the coin. Insert the screwdriver tip into the soil above the coin at a 45° angle. Insure that the tip is about halfway between the surface and the item itself. Lift the grass up a couple of inches and probe for the coin. You should be able to work the coin up with the screwdriver without too much damage to the grass. Lift the coin from the hole, pack the ground back down, and then step on it. This way you will not tear up the lawn and leave dark spots.

A small digging tool or spade can be used to uncover large items but the same rule of filling holes and removing trash still applies.

Currently, it is unlawful to use a metal detector in national parks or within the boundaries of national monuments, state parks, and designated historical sites. So keep clear of these areas. Violators have not only been fined, but have had their detectors, automobiles, and other belongings confiscated as well. If you are not sure of existing city ordinances regarding treasure hunting in your area, contact your local officials.

RETURNING TO A LOCATION

Once you have located an area which turns up some interesting—if not valuable—treasures, you probably will want to spend several days combing over the site. How does one mark a spot for his return without making it obvious for someone else to find the location?

Let's assume you are searching a 200-acre Civil War battleground and just about dark you locate a small area where several unfired cartridges are found. Let's suppose you also find melted lead with the metal detector so you naturally assume that this particular spot was probably a camp. You want to keep digging but darkness is moving in fast. Other treasure hunters are all around you so you don't want to make the spot too obvious. Therefore, you cover the hole with the fresh dig and then cover this with dead leaves, making them look as natural as possible. Now you must mark the spot. First pick some object—the south side of a large oak tree, a corner of a building, any permanent object. Line this object up with another object farther away. Insure that this imaginary straight line crosses over the treasure spot. Sighting along a straightedge is probably the best way to do this. Now turn and locate two other objects and repeat the procedure. It's best if the two imaginary lines intersect over the spot at right angles or nearly so. If this is accurately done, then you have "marked" the exact location of your find. You will, of course, want to make a note of these objects so you don't forget them.

If you carry a compass along with you, the marking of these spots can be done more accurately. First, while you're standing on the spot, locate an object in the distance. Determine the bearing of the object with your compass. Next, locate another object in the distance and find its bearing also. The bearings will intersect at the treasure spot. Again, it's best if the bearings intersect at about a 90° angle. Make a note of the bearings. When you return to the spot later, find one of the bearings with your compass. Walk along this bearing until you find the spot where the other bearing to the other object intersects.

It is difficult to remember more than a few of these spots so a record must be kept. A rough sketch with notes is probably the best method of keeping records of your previous finds. For this, a small notebook and pencil comes in handy. By using the notebook for your treasure hunting records, you can return to the better spots year after year.

Many treasure hunters purchase a small battery-operated tape recorder that may be carried in a shirt pocket. At any time you can flip a switch and record data.

Various maps, either purchased or self-made, may sometimes help to illustrate an area better than notes. Maps of every description may be purchased from the U.S. Geological Survey. In addition to selling over-the-counter maps throughout the United States, this agency distributes a free leaflet containing a dealership list and instructions in map-reading.

The U.S. Geological Survey produces and sells a series of state base maps and topographic maps in three scales—1 inch= 4 miles, 1 inch = 1 mile, and 1 inch = 2000 feet. The maps sell for approximately 50¢ each. When ordering, specifically identify your geographic areas of interest. Write to: Geological Survey, U.S. Department of the Interior, Washington, D.C. 20242.

The Department of Agriculture supplies aerial photographs for every region of the United States and its territories. Different scales are available from 1 inch = 400 feet to 1 inch = 1667 feet. The photographs range in price from about $1 to over $10 each. To obtain an aerial photograph, outline, on a road map, the section you wish to have and request photographs that cover this particular section. Be certain to mention the county name and other information that will help the agency identify the area in question.

For aerial photos east of the Mississippi River, write Eastern Laboratory, Aerial Photography Division, ASCS-USDA, 45 South French Broad Avenue, Asheville, North Carolina 28801. Areas west of the Mississippi are serviced at Western Laboratory, Aerial Photography Division, ASCS-USDA, 2505 Parley's Way, Salt Lake City, Utah 84109.

Another source for detailed maps is your state highway department. Most states have county maps which are either free of charge or else sell for only a small fee. These maps usually show all state-maintained roads along with streams, rivers, and other landmarks. Contact your nearest state or county highway office for further information.

Sometimes you will find it necessary to make your own maps in order to help you remember where you have found treasures. Before you start to map an area, make certain that you pick a suitable place on the paper for your starting point so you will be able to get all the desired information on the paper.

Map-making consists of three basic factors: direction, distance, and details. Your first step is to mark magnetic north on the map. This helps to properly orient other markings. Any other directions are determined with compass bearings. You can measure distances on the ground by stepping them off. You can then draw the distances on the map to some predetermined scale. The scale, of course, will vary according to the amount of detail required on the map, but a good scale to begin with is 40 steps equals one inch on your paper.

This will give your map an approximately one inch to 100 feet scale. You will need a small ruler to mark off the points on the map.

Landmarks and other details are shown on the map by symbols; you can develop your own to indicate such finds as Civil War relics, Indian relics, coins, old bottles, etc.

One of the most accurate ways to make your own maps is with the use of a pantograph. This mechanical device can be used for copying a map or diagram in the same scale, a reduced scale, or an enlarged scale. The degree of enlargement or reduction is determined by settings on the instrument.

Here's an example of how you can use the pantograph. Assume that you plan to treasure hunt along a creek bed where you know a Civil War skirmish took place. Let's say this skirmish—according to an old war map—took place in a 600-square yard area which includes the creek bed and land on both sides of it. Knowing that you want to record data about this area, you decide to map the area. You have a county map obtained from your state highway department, but you want to show the locations of your finds more exactly than the map allows.

Begin by making an enlargement of the county map by tracing it with a pantograph. Make notes of any prominent landmarks. Then take your enlargement into the field. When a find is located, use your compass to get bearings and then pace off the distance to each find from some landmark. All of this data is then recorded on your enlarged map.

All successful treasure hunters make mental notes whether they realize it or not, but it has been said that "the strongest mind is weaker than the palest ink." Those observations that are left to memory often become hazy with time; however, if a written or taped record is made—along with notes marked on a suitable map—the information can be preserved indefinitely.

Chapter 5
Treasure Hunting In Water

When the words "underwater treasure" are mentioned one normally thinks in terms of sunken Spanish galleons loaded with silver bullion worth millions of dollars. Those few persons who search for these galleons, however, will quickly tell you that an enormous investment is required to undertake such expeditions. Such an investment is impossible for the average hobbyist. Even when finds are made, very seldom is the worth of it even enough to cover expenses.

Fortunately, there are treasures found underwater that can be within the reach—financially and timewise—of almost everyone. You probably won't get rich overnight but the fun and small profits derived from such an undertaking will certainly make the practice worthwhile. And you don't even have to live near the coasts; there's treasure right in that river or lake near your home. Perhaps the treasure will consist of a fisherman's tackle box, an outboard motor, or a camera. Or maybe the treasure will be gold panned from a mountain stream. Many families all over the country are bringing home up to $500 a weekend by gold-panning.

An if you like skindiving there are dozens of lost ships on the bottom of the Great Lakes waiting to be explored. This kind of treasure hunting doesn't have to be expensive but it does take skill.

LOOKING FOR GOLD UNDERWATER

Gold, the universally accepted symbol of wealth, is found everywhere, but some people consider only streams in the West as gold-bearing. But did you know that in the decades before the California Gold Rush of 1849 the most productive gold-bearing areas in the United States were located in Virginia and Georgia? So regardless of where you may be located chances are that you can find a stream or river nearby that will yield some gold dust and perhaps a gold nugget.

It is estimated that there is more gold left in the ground and crevices of streams than all the gold that has been taken out. Elaborate mining equipment is not needed to recover a good portion of this gold. All that is needed is a basic knowledge of the better spots to locate gold, an inexpensive gold pan, and some practice in using it.

Beginners who have never panned gold can gain some skill by practicing with birdshot (about the size of BBs). Put the birdshot in the pan with gravel, dirt, and sand. Then use any water container—such as a wash tub—for your experimenting. When you have gained enough experience with the birdshot you might as well get started on the real thing.

Select a stream and place gravel, sand, etc., in your gold pan. Then place pan and all underwater. Use your hands to break up clay or mud and to remove the larger rocks. Let the small gravel and sand sift through your fingers. Keep the pan underwater at all times.

With the pan still underwater, oscillate the pan so its contents swirl causing the gravel and sand to "loosen." This loosening will let the heavier gold, minerals, etc., settle to the bottom. As you oscillate the pan tilt its forward rim down, letting dirty water, sand, and gravel gradually wash over the edge. Continue this circular swirling motion, pausing occasionally to rake the cleaned material back into the water. If you are careful to keep the pan underwater at all times materials like gold, lead, etc., will begin settling to the bottom of the pan. And if you're using a pan with built-in riffles, they will do the rest to hold this gold intact.

Start using a gentle, but firm, side-to-side motion and let the lighter material on top gradually slip over the edge of the pan, still keeping the pan under water. Tilt the edge of the pan back toward you intermittently. This helps the heavier concentrates to gather in the built-in riffles, or gold traps.

The heavier concentrates should now be caught in the riffles but continue the side-to-side motion a while longer, letting the lightest material slip over the edge. When the remaining material is down to about one teaspoon stop and allow a very small amount of water to enter the pan. Then carefully spread the concentrates and inspect the remaining black sand for nuggets, specks of gold, garnets, sapphires, and other precious metals or minerals.

One of the biggest problems facing the amateur prospector is separating the very fine particles of gold from the panned gravel and sand. The small particles of gold drift in this debris, making the retrieval of them very difficult. When the weekend prospector has to spend endless hours of tedious labor trying to separate a few gold flakes from thousands of grains of worthless black sand it may seem best to toss the entire mess back into the stream and forget it. However, with a few simple materials—and a little know-how—that precious gold can be saved by a process known as amalgamation.

Besides your gold pan, you will need a small magnet, a few drops of mercury, and a chamois cloth. First, carefully pan out most of the large waste particles so that all that remains is black sand and other fine material that could contain flakes of gold. Remove most of the black sand with the magnet; the black sand will stick to the magnet, leaving a nonmagnetic material in the pan. This material is called amalgam. If there are any gold specks they will be in the amalgam.

Place a few drops of mercury in the pan with the amalgam and mix the contents underwater. The mercury should be worked thoroughly through the amalgam. Any gold in the amalgam will adhere to the mercury. The ball of mercury will appear rough on the outside after it has gathered the gold. At this point the remaining wastes in the pan should be panned out, leaving the ball of mercury in the pan.

Then slide the ball of mercury out of the pan onto the piece of chamois cloth. Strain the mercury through the cloth; twist the cloth hard. The mercury will be forced through the cloth while the gold and some of the mercury will remain in the chamois cloth. Heat what remains in the chamois cloth over a fire. This burns off what mercury is left and leaves only gold.

Warning

Do not inhale the fumes or smoke. They are poisonous.

Instead of heating the amalgam over a fire, you can cut a potato in half and scoop out a small cavity in one of the halves. Pour the amalgam into this cavity, place the two halves back together, and wrap them in several layers of aluminum foil. Bake this potato for approximately one hour then remove the foil. You will find that the heat has driven the mercury deep into the potato while the gold remains in the cavity. Don't eat the potato—the mercury has made it poisonous.

FISHING WITH A MAGNET

If you don't care to get your feet wet you can purchase a PVC coated magnet for under $20. Tie a cord to it and drop it over the side of your boat to recover fishing tackle, outboard motors, anchors, and all kinds of ferrous metallic objects. You can hunt the bottoms of lakes, rivers, bays, and even oceans. A five-pound magnet will lift up to 150 pounds (more underwater), depending on the surface of the object it is lifting. The greater the gripping surface, the more effective the magnet.

Boat propellers are always being lost in waters. Each of these recovered with your magnet could be worth $20 (the price paid for your magnet) or more. And it doesn't take long to find a buyer. For every propeller you pull up, it's like pulling up $20.

Covering a large underwater area can be accomplished by "trolling"; that is, by dragging your magnets behind the boat as the boat is moved slowly forward. Continue this procedure back and forth across the search area until the entire bottom is combed.

FISHING WITH A METAL DETECTOR

An electronic metal detector can be used to locate both ferrous and nonferrous metals and minerals. Garrett Electronics of Garland, Texas, manufactures a BFO (beat frequency oscillator) detector with optional underwater search coils in either 5-inch or 12-inch diameter sizes. Both sizes can be lowered down into 50 feet of water to detecto objects on the bottom or just beneath the sand and mud. It will locate boats, motors, fishing tackle, guns, and other objects.

When using the detector, however, you will have to dive for the objects rather than retrieve them with a magnet tied to a cord. A diver can maneuver the coil over the bottom while the operator in a boat monitors the control housing. A tug on the cable will alert the diver to a find. Scuba equipment will be necessary for the deeper waters but a snorkel and mask should suffice for shallower depths.

Fig. 5-1. A group of divers about to begin an underwater expedition in an inland river.

You can find more kinds of treasure with an underwater metal detector than with a magnet. One treasure hunter, diving off a public beach, recovered several dollars in small change, three gold rings, several gold earrings, and other valuables in less than one hour with his underwater metal detector. You can do the same off a public beach near your own home. A public lake or stream will offer the same chances; even the bottom of a swimming pool might offer surprises.

DIVING FOR TREASURE

Since the introduction of skin and scuba diving a few decades ago, many amateur treasure hunters have been able to hunt for many of the sunken treasures that were once only open to costly professional expeditions. Inland rivers and lakes also offer the amateur diver territory to search for sunken treasure (Fig. 5-1). One recent find was that of several Civil War cannons discovered in a North Carolina river by a diving party. The find was valued at over $100,000!

Besides the many treasure-laden Spanish galleons that we hear so much about, there are hundreds of other lost ships off both coasts (Fig. 5-2). For example, during the Revolutionary War, more than 500 ships were lost off the east coast of the United States between Maine and Florida. Only a very small percentage of these ships carried great amounts of treasure but all of them sank with items which are now of great value. Some of these items include cannon, swords, firearms, tooks, glass bottles, and hardware of every description.

While very few of these ships (under 20) are known to have carried any substantial amount of wealth, all of them certainly carried some coins when they went down. Any of these coins would be

Fig. 5-2. A diver combing the bottom of a cove for a sunken Spanish galleon.

Fig. 5-3. A diver searching for the remains of a ship.

worth a great deal of money in this day and time. In fact, some coins from this period have brought over $100,000 each. Others have sold for $5,000 and up!

The waters off the coast of California are generally deeper and colder than those off the East Coast but there's still plenty of treasure to be found on sunken ships in this area (Fig. 5-3). A typical find might include old coins, firearms, gold nuggets, and jewelry. Silver and pewter objects as well as clay smoking pipes are also relatively common around sites of the older wrecks.

I certainly don't want to make diving for treasure sound easy; it certainly is not. It's a lot of hard work. But thousands of amateur divers are finding mini-fortunes nearly every time they go out. These include Indian relics, prehistoric animal bones, gold jewelry, coins, war relics, old bottles, human skulls, and the like. Although searching for these treasures may not be as profitable as finding a treasure galleon, the search can be just as exciting and rewarding.

LAWS GOVERNING UNDERWATER FINDS

Don't forget that anything found underwater has (or did have) an owner and you can't necessarily claim it merely because you found it. In nearly all significant finds the federal and state governments will have certain claims on it, perhaps giving the finder a small honorarium for his trouble. In some cases they may claim the whole treasure.

It is possible to buy salvage rights on some sunken ships from insurance companies. This is because the insurance company that paid the claim to the original owner of the sunken ship becomes the rightful owner of the ship. Many divers who find such wrecks examine the remains, make an estimate of the salvaged value, and then make the insurance company an offer.

Therefore, before you start removing an underwater discovery always check with local, state, and federal authorities to determine what laws pertain to your find.

BOATS FOR TREASURE HUNTING

If you're going to be doing a lot of treasure hunting in lakes, rivers, and oceans, you'll need a boat. But what kind of boat? The expense of a boat for *occasional* underwater treasure hunting could only be justified if the craft can be used by the whole family for pleasure-fishing, cruising, etc.

There are at least two dozen basic types of craft to choose from so there is really no reason for not having the boat that is best suited to your individual needs—both as a treasure-hunting vessel and as a pleasure boat.

What then are the basic steps in finding the right boat for you and your family? First, and most important, is to decide upon the type of waters the boat will be used in. Will it be used mostly in bays and coastal waters? For trips on inland rivers and streams? For a cruise to one of the tropical islands? The boat's primary use should be decided first. Then begin thinking about other activities—in the order of their importance—which the boat may be called upon to participate in. Next,decide upon the number of people you expect to carry in the boat and how the boat will be transported.

And then decide what types of treasure you will be searching for. It stands to reason that if you're looking for old cannon weighing a few tons, you won't be able to transport them in a 14-foot rowboat. When all these decisions have been settled you will have a sound basis for selecting a boat that will come pretty close to fulfilling all your boating needs.

Many people who select their first boat often overlook dozens of things that should be considered before making a purchase. For example, I knew one person who bought a boat to use in the waters near his home. But he forgot to consider the *depth* of the waters. Now he has a $3500 boat that is forever being stuck on sand bars! And not long ago a couple invested a couple hundred dollars in a metal boat to use near their beach home. But in only a few short months their craft was full of holes. Salt water.

These and other mishaps can be avoided if you acquire a basic knowledge of boats in general and then find out the advantages and disadvantages of each type. Begin by reading through as many boating magazines as you can get your hands on. Attend boat shows and listen to what the manufacturers' representatives have to say. Talk to some of the "old salts" down at the basin. Observe other boats in operation and try to talk to people who actually use boats in treasure-hunting activities. All of these things can help you make the right decision when it comes to selecting your own boat.

To further assist you in your selection, the descriptions that follow will give you a brief introduction to some types of boats that are particularly well suited to the weekend treasure hunter.

The *flatboat* is known by various names: john, jon, joe boat, punt, gravey, scow, and batteau. This type of craft is simple in design yet durable and safe. Most flatboats are constructed mainly out of yellow pine and plywood and are quite suitable for treasure hunting on inland waters. They will also serve the owner satisfactorily for fishing, hunting, or trapping in streams, bayous, or lakes. You can power this type of boat with a small outboard motor. Or you can row it or pole it or paddle it as you choose. If designed properly, it will ride high in the water to offer less resistance than many other types of boats.

Many flatboats are lightweight which makes them ideal for transporting in the bed of a pickup truck; two men can easily carry them. Their durable undersides make them the perfect boats for use on rivers and streams where there are protruding submerged rocks.

Most flatboats can also take a considerable load. Most can easily carry you, your equipment, and several hundred pounds of treasure.

A variation of the flatboat is the *pirogue*, a flat-bottomed double-ended craft used by the natives of Louisiana. It's a craft used mostly on protected waters—especially if much shallow water is encountered. The average pirogue is approximately 14 feet long. It's lightweight and easy to build from practically any type of wood. The design has proved to be satisfactory for many decades and if your treasure-hunting activities take you to shallow inland streams this might just be the craft for you.

If you plan to do most of your diving in lakes, bays, or coves where the waters are normally calm you might want to use a *pontoon deck* boat. Such boats resemble a floating patio. They provide excellent porch-like space for storing diving gear, finds, and the like. There's also plenty of room for a charcoal broiler so you can have fish fried aboard your floating patio after a day of diving for treasure. If you really want to go whole hog, add a cabin to the platform to make it suitable for camping and extended cruises on lakes and other protected waters.

Another type of inexpensive craft that is used extensively by offshore divers is the *dinghy*. The dinghy was originally used as a means of getting to and from larger vessels but in recent years they have proven to be very handy when equipped with a sail, centerboard, and rudder. In fact, it is not unusual to see the owners of luxury yachts neglecting their "pride and joy" for the fun of sailing their yacht's dinghies.

The weekend treasure hunter will find that a good dinghy will quickly become close to his heart. It can get you to treasure-hunting waters by sail, oar, or small outboard motor. It's both easy and inexpensive to build, has a multitude of practical uses, and is easily transported on the top of the family car. It makes a good base when skindiving for sunken treasure. For a first boat to be used on lakes, bays, and other protected waters, the little 8-to 12-foot dinghies are hard to beat.

The traditional *canoe* is ideal for quietly fishing and exploring the shallow shorelines. Most canoes are very lightweight and can be

Fig. 5-4. A modified version of the kayak. (Courtesy Glen L. Marine Designs.)

handled and carried by one man. Their only drawbacks are that they're easily damaged and will tip over easier than most other craft.

Many people believe that the *kayak*—both the original Eskimo version and many of its modified forms—is still the best all-around one-man boat for nearly any type of activity on water. When equipped with watertight bulkheads it can be used along coastal waters as well as protected inland waters. Most newer versions have room for gear and finds (Fig. 5-4). The average kayak is considered to be more stable than a canoe but is just as light and as easy to transport. Some varieties even come in a takedown folding type of design.

The traditional way to power a kayak is with a double-ended paddle. But with a special side bracket a kayak can be powered with a small outboard motor. Many kayaks have been rigged for sailing.

You should select a *cabin cruiser* only after you carefully consider the load the craft is expected to carry, the distance you plan to cruise, and how you plan to use the cruiser (Fig. 5-5). Almost all cabin cruisers have facilities for cooking and sleeping and they usually have ample cockpit space for lounging and eating. The cost of any type of cabin cruiser is going to take a big hunk out of your bank account. But such a craft can give you and your family a lot of pleasure; it'll also enable you to cover a lot of treasure-hunting waters.

If you want to enjoy all the comforts of home while you're treasure hunting, then a cruising houseboat is for you (Fig. 5-6). Such houseboats have lots of room for sun bathing or fishing on the decks. Inside the cabin you'll find plenty of headroom, provisions to sleep four or more adults and a private enclosed toilet room. The galley will usually have plenty of room for the latest marine cooking equipment plus storage space. Houseboats *under* 25 feet are usually trailerable so take this into consideration if you're planning to go this route.

If you want higher performance with fewer frills, consider the large variety of *utility boats*. (A typical utility boat is shown in Fig. 5-7.) Such boats are all-purpose crafts for use in bays and coastal waters. They are recognized by their large open cockpits. They can be used in fishing, skindiving, camping, and towing skiers.

Despite their open cockpits, many utility boats have raised decks that provide high freeboards forward and space for berths below deck. These boats can also take plenty of power so that high speeds are possible. Yet their compact size makes them ideal for trailering and launching. Such boats can be used for overnight camping, cruising, sport fishing, water skiing, and just about any other water sport activity.

Fig. 5-5. Cabin cruisers normally have facilities for cooking and sleeping.

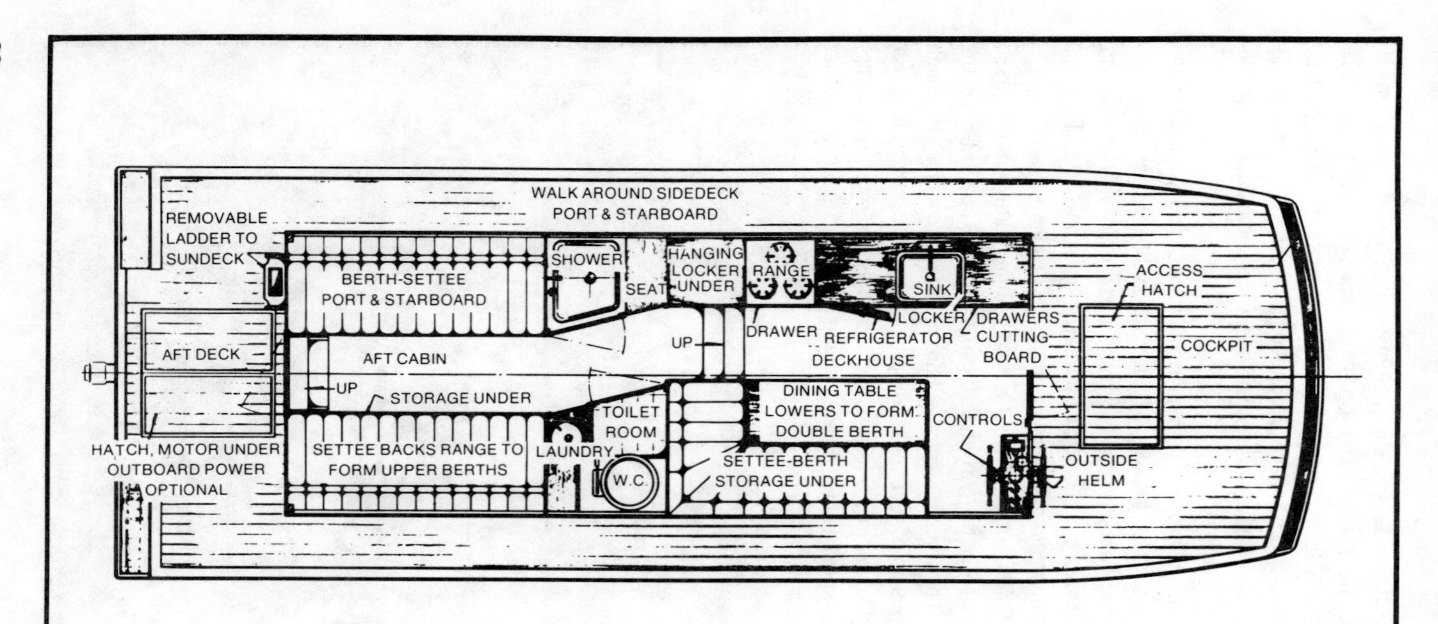

WALK AROUND SIDEDECK PORT & STARBOARD
REMOVABLE LADDER TO SUNDECK
BERTH-SETTEE PORT & STARBOARD
SHOWER
SEAT
HANGING LOCKER UNDER
RANGE
SINK
ACCESS HATCH
AFT DECK
AFT CABIN
UP
DRAWER
REFRIGERATOR
LOCKER
DRAWERS
CUTTING BOARD
DECKHOUSE
COCKPIT
UP
STORAGE UNDER
TOILET ROOM
DINING TABLE LOWERS TO FORM DOUBLE BERTH
CONTROLS
HATCH, MOTOR UNDER OUTBOARD POWER OPTIONAL
SETTEE BACKS RANGE TO FORM UPPER BERTHS
LAUNDRY
W.C.
SETTEE-BERTH STORAGE UNDER
OUTSIDE HELM

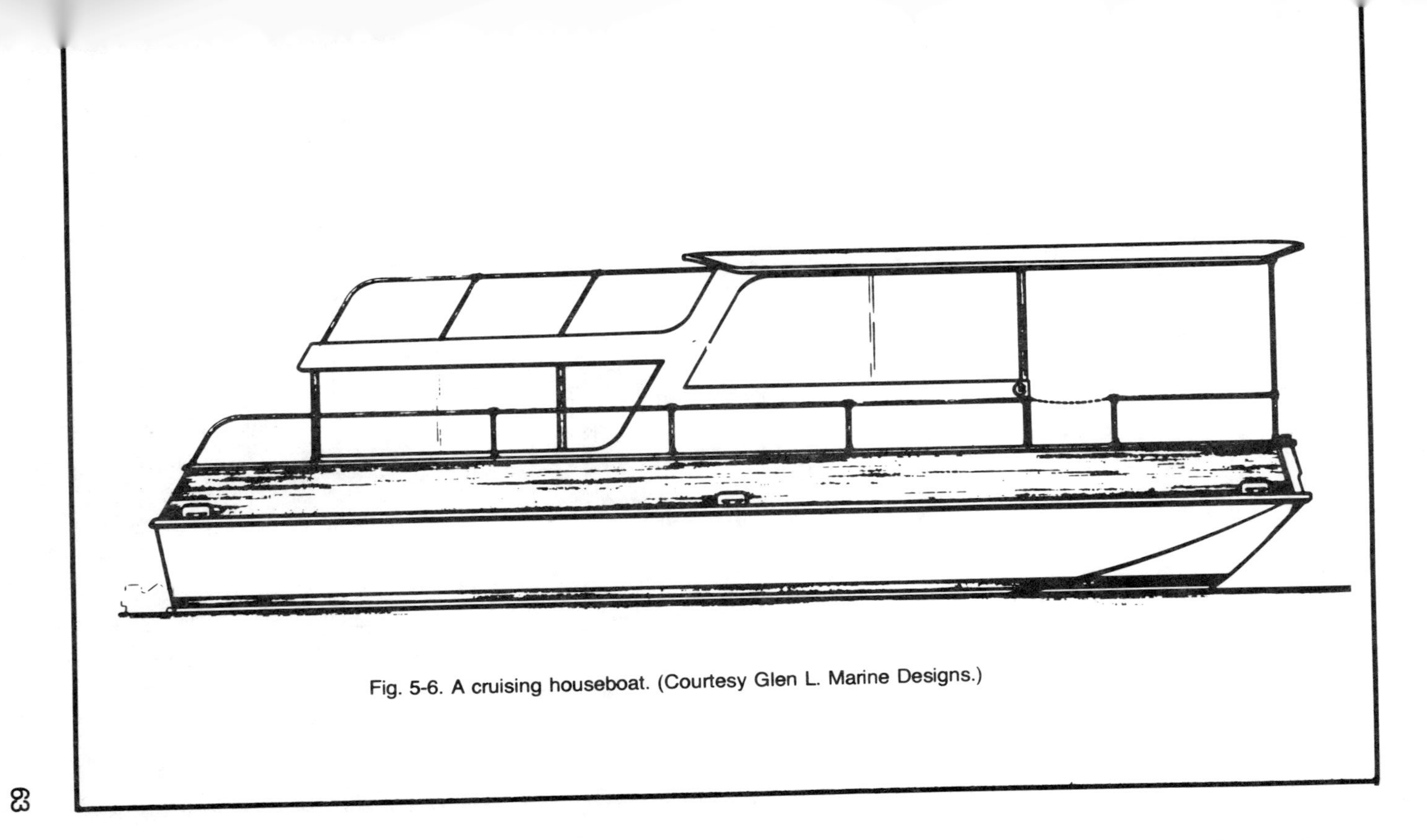

Fig. 5-6. A cruising houseboat. (Courtesy Glen L. Marine Designs.)

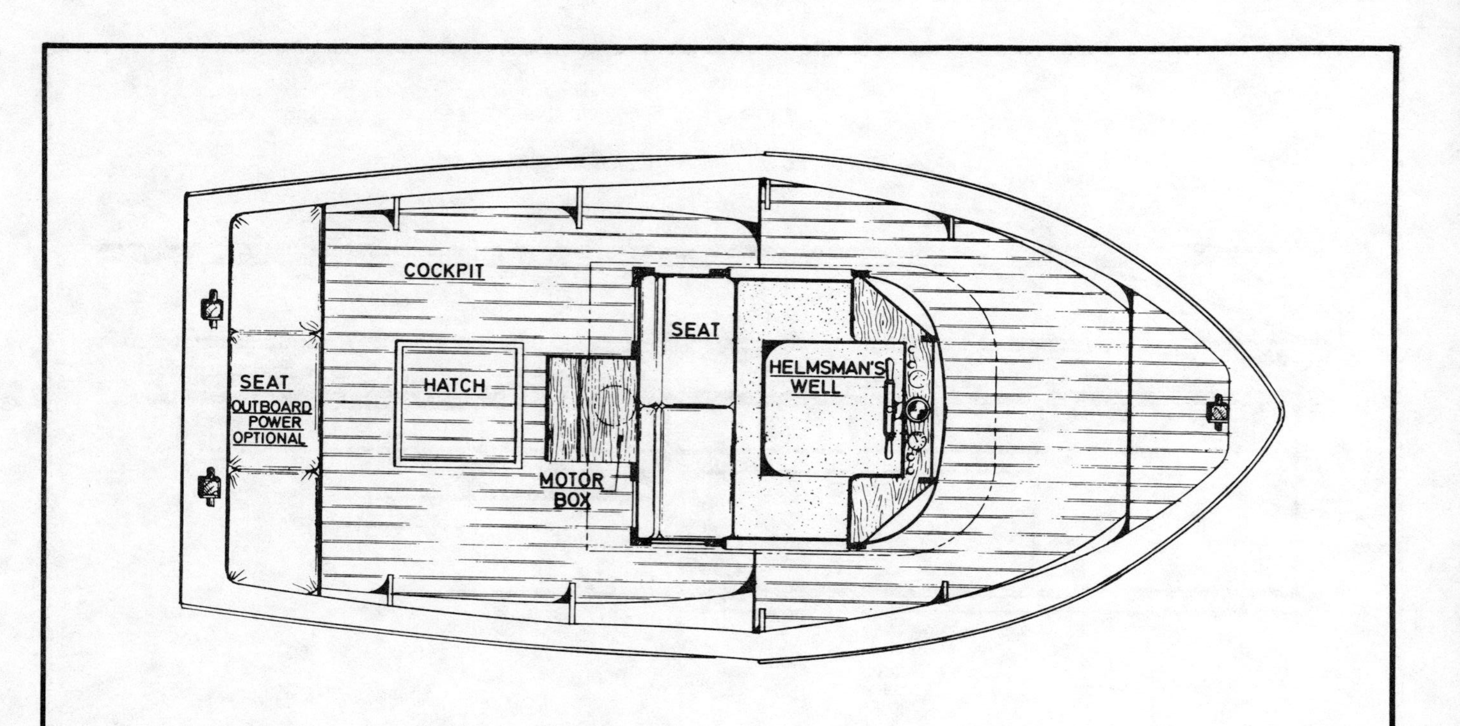
COCKPIT
SEAT
HELMSMAN'S WELL
SEAT
OUTBOARD POWER OPTIONAL
HATCH
MOTOR BOX

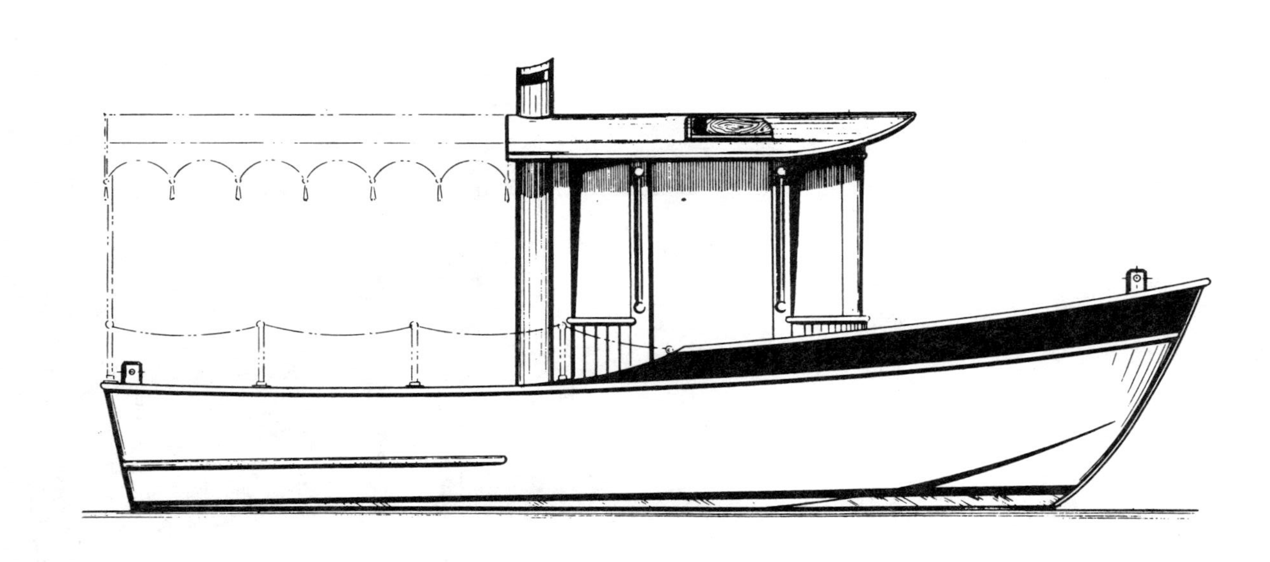

Fig. 5-7. This utility boat is stable enough for treasure hunting in coastal waters. It has plenty of room for gear and treasures. (Courtesy Glen L. Marine Designs.)

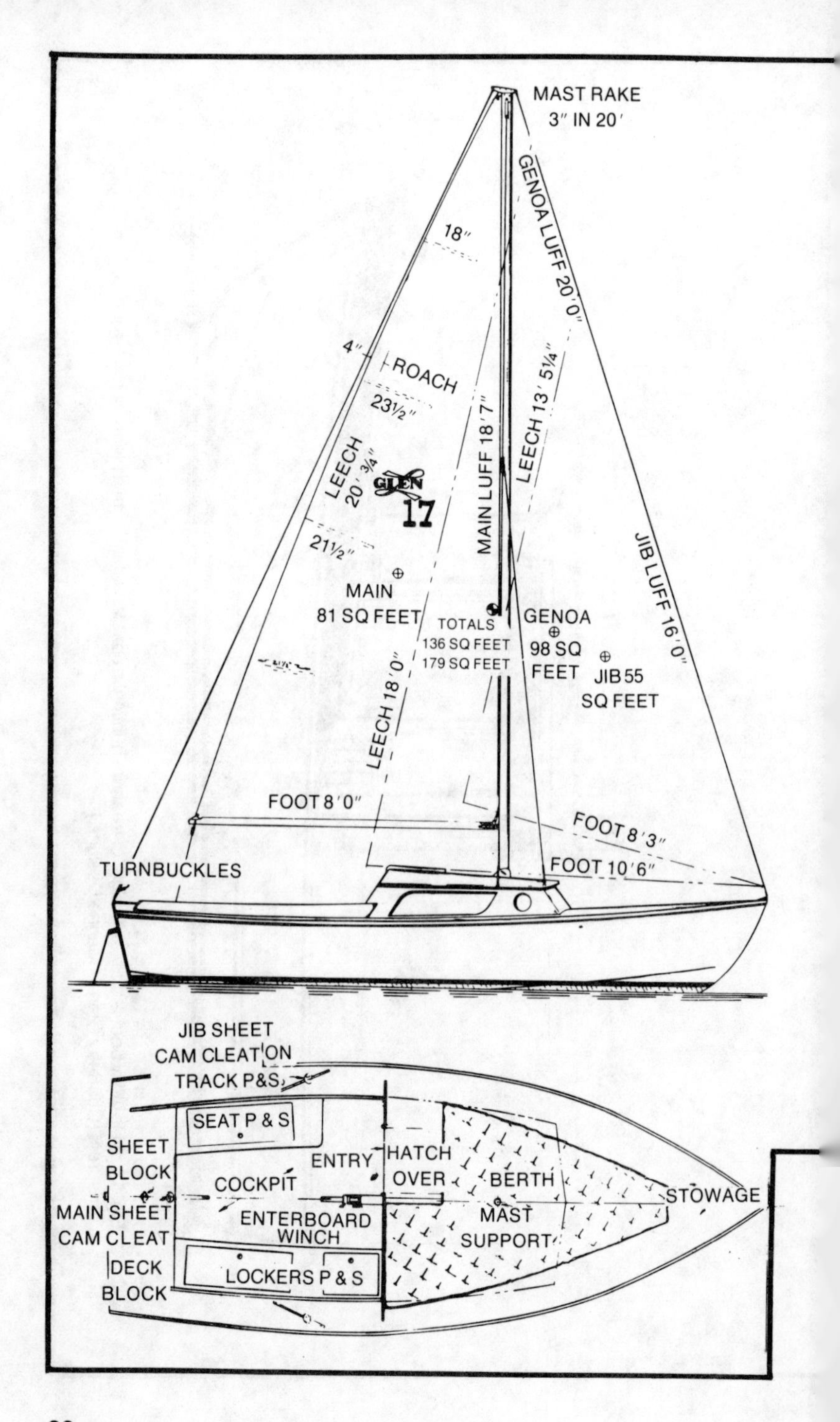
MAST RAKE
3″ IN 20′
GENOA LUFF 20′0″
18″
4″
ROACH
23½″
LEECH
20′ 3/4″
GLEN
17
MAIN LUFF 18′7″
LEECH 13′ 5¼″
JIB LUFF 16′0″
21½″
MAIN
81 SQ FEET
TOTALS
136 SQ FEET
179 SQ FEET
GENOA
98 SQ
FEET
JIB 55
SQ FEET
LEECH 18′0″
FOOT 8′0″
FOOT 8′3″
FOOT 10′6″
TURNBUCKLES
JIB SHEET
CAM CLEAT ON
TRACK P&S
SEAT P & S
SHEET
BLOCK
ENTRY
HATCH
OVER
BERTH
COCKPIT
STOWAGE
MAIN SHEET
CAM CLEAT
ENTERBOARD
WINCH
MAST
SUPPORT
DECK
BLOCK
LOCKERS P & S

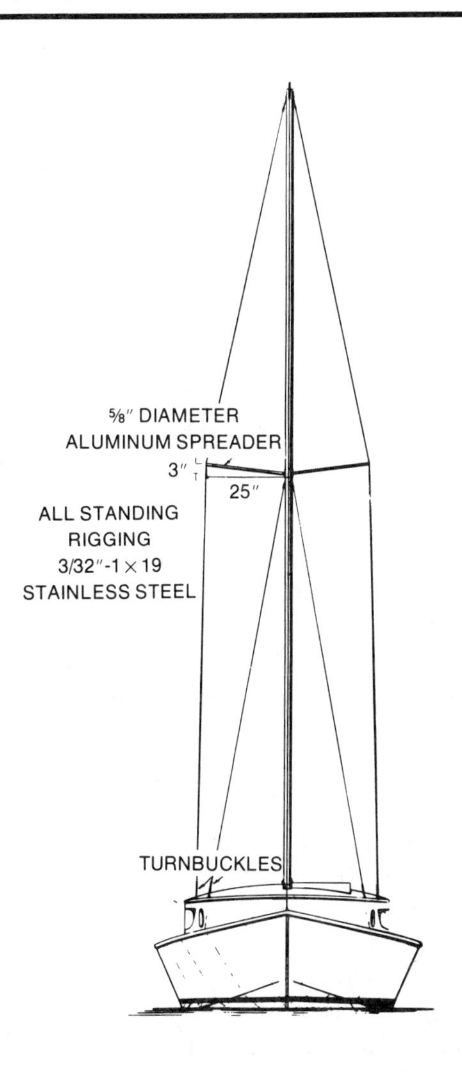

Fig. 5-8. A typical sailing cruiser. (Courtesy Glen L. Marine Designs.)

There are many medium-sized, earily transported *sailing cruisers* designed for single-handed sailing (Fig. 5-8). Sleeping facilities are provided for two or more people—with space for a folding toilet or even a full-sized head. The cockpit on the smaller rigs provides room for a portable ice box, stove, auxiliary outboard motor, and sail bin.

Chapter 6
Equipment For Treasure Hunting In Water

The types of equipment for treasure hunting in water are nearly as varied as the treasures themselves. You can use an inexpensive pie tin to pan for gold in a nearby stream. Or you can don some fancy scuba gear to search out the deep. You can buy or you can make your own. The important thing is that your equipment gets the job done.

GOLD HUNTING GEAR

If you enjoy the great outdoors, spectacular scenery, and healthy, invigorating activity, then prospecting and dredging for gold is for you. Picture you and your family camping near a cool mountain brook. Imagine yourself doing a little fishing, a little napping—and leisurely panning $100 worth of gold dust.

It is easy to see why prospecting for gold has proven to be one of the fastest growing hobbies today. And fortunately many an amateur prospector has more than paid for his equipment in a single season and some have found some real bonanzas.

The modern day hobbyist or part-time prospector will probably want to start out with one of the gold panning kits that sell for less than $10. Most of these kits include a gold location map covering the entire United States, panning instructions, gold samples, panning sand (for practice), and a plastic pan with built-in riffles and positive gold trap to make the panning easier. Such a kit is available from Keene Engineering, Inc., 9330 Corbin Ave., Northridge, California, 91324. At this writing, the kits sell for $5.95 plus tax, postage, etc.

If you want to go a step further, this same firm also offers a more sophisticated kit for under $50. This kit contains a mini-sluice, a prospector's pick, a plastic gold pan, a gold magnet, a sample bottle, a prospector's shovel, a crevice tool, a magnifier, and a reference book. The mini-sluice is about three feet long and 10 inches wide and is lined with removable riffles, a diamond expanded screen, and a matting. Merely set it in the water, shovel gravel into the hopper section, and let it do the work for you. The magnet is used to remove the hard-to-get-rid-of black sand in the bottom of your gold pan.

A surface dredge consists primarily of a portable sluice box with a vacuum hose attached for dredging gravel underwater (Fig. 6-1). Attached to the suction hose is a vacuum device commonly referred to as a suction jet, or eductor. Water is pumped to the jet under high pressure, creating a powerful vacuum (Fig. 6-2).

As the material is vacuumed into the sluice box, it first enters the hopper baffle. This hopper baffle causes the material to reverse its direction of flow for classification over a trommel screen. The larger gravel passes *over* the screen while the smaller material washes *through* the screen for a more selective separation in an area of lower velocity. The smaller material then enters the riffle section, causing any solids yet in suspension to settle and separate by weight or specific gravity. This process has proven to be the most effective means of recovery in a portable sluice. Beneath the riffle section is a matting that seals the riffles and aids in holding fine concentrates (gold, lead, etc). The riffle section is easily removed by quick release latches, leaving the concentrates exposed on the removable matting. The concentrates can then be panned as described previously. The dredge can be operated for several hours without cleaning. All components are made of either lightweight durable aluminum or steel.

A bulb sniffer is sometimes used underwater for extracting small gold specimens. A bulb sniffer can be just a common kitchen baster. When its bulb is squeezed and released, suction is created and sand and gold dust are sucked into the sniffer's tube.

If you want to do your own appraising of your find, you can purchase a metal test kit for appraising and testing gold, silver, platinum, brass, nickel, and German silver. These test kits contain three acid bottles with ground-glass stoppers and applicators, a test

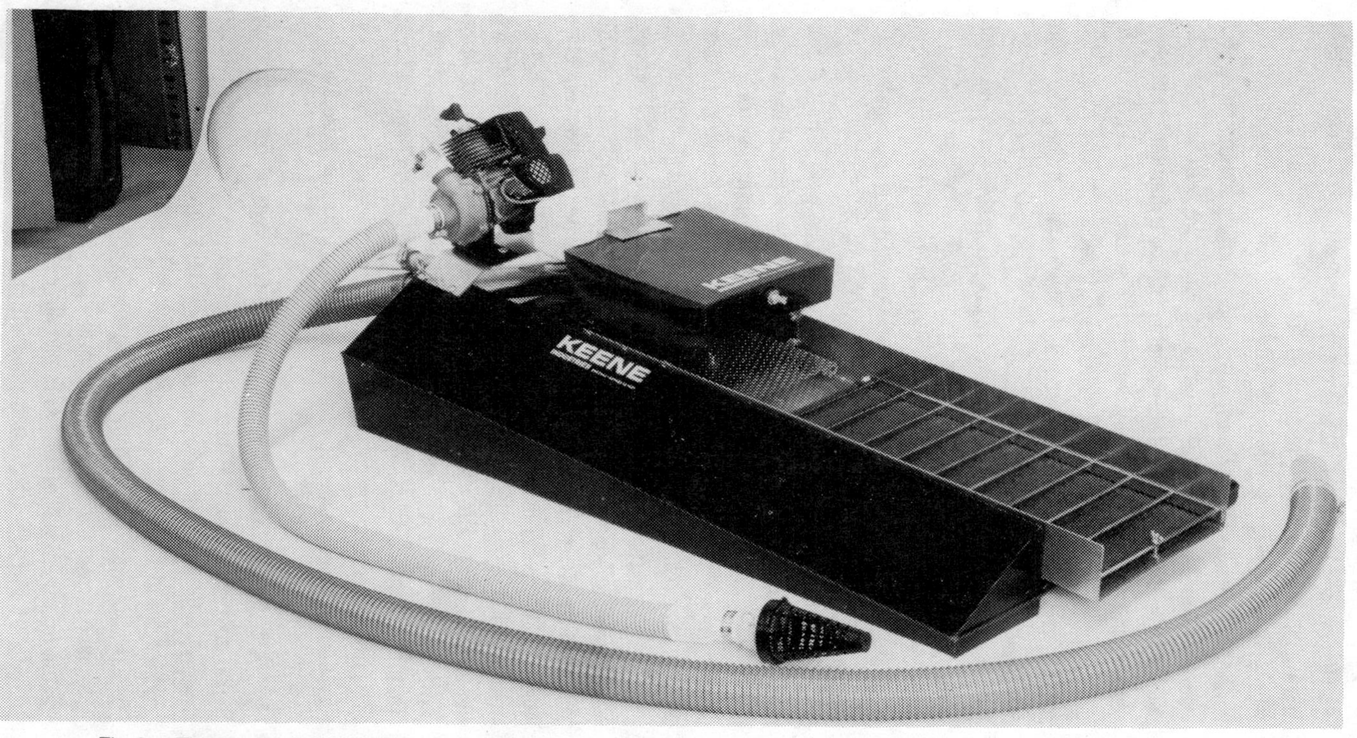

Fig. 6-1. This highly sophisticated dredging machine is capable of moving up to 3 cubic yards of loosepacked gravel per hours. (Courtesy Keene Engineering, Inc.)

stone, and a set of gold testing needles. Normally, the acids must be purchased separately. You will also need a balance for weighing gold and a set of metric weights from 10 milligrams to 100 grams.

In recent years, some gold prospectors began using scuba gear for underwater prospecting and it has turned out to be very productive. Most prospectors use only a mask and a snorkel for their operations, but some go the whole route; that is, complete wet suit,

Fig. 6-2. A prospector using a dredging machine.

knee protectors, diving hook, warm neoprene boots, air tanks, and the rest. This same gear is used for diving for treasure of all types in both fresh and salt waters.

MAGNETS

Many "dry" divers have been searching the bottoms of bays, rivers, lakes, and even oceans without getting so much as their feet wet. They have been tying a line to a lightweight magnet and dropping it over the side of their boat to "catch" metal objects of all kinds. Edmund Scientific Company offers several magnets which are suitable for this purpose. They sell for from $9 to nearly $20—with magnetic pull from 25 pounds to over 150 pounds. These will, of course, lift more weight in water.

DEPTH SOUNDERS

For locating sunken treasures underwater, a depth sounder can improve your chances tremendously. One type (unassembled) is available from Heathkit; it is a recorder type that provides a continuous indication of water depth and the depth of submerged objects. The recording accessory makes this instrument extremely valuable as a navigational aid if you must follow coast lines or operate in harbors, in fog or at night. The high sensitivity and the straight black-line marking on white dry-lined paper make it possible to locate even schools of small fish, and, in most cases, individual large fish.

The beam pattern is typically shaped like a tear drop and extends in all directions from the axis of the transducer. Figure 6-3 shows a cross section of the beam pattern (at three sensitivity settings). The unit is only able to record objects *within* the beam pattern. Therefore, at low sensitivity setting A, only fish #1 would show on the chart. Note that fish #2 at the same depth as fish #1 would not be detected. At a medium sensitivity setting, fish #1 and #3 would be detected. At a high sensitivity setting, all fish (#1, #2, #3, and #4) would show on the chart. To "see" underwater objects, it is necessary to set the sensitivity higher than is required to detect the bottom. This is because the reflected signal from small objects is much weaker than the reflected signal from the large bottom area.

Many lakes and waterways have already been surveyed by the government. Publications containing information about these lakes and waterways are available at a nominal cost from the Government

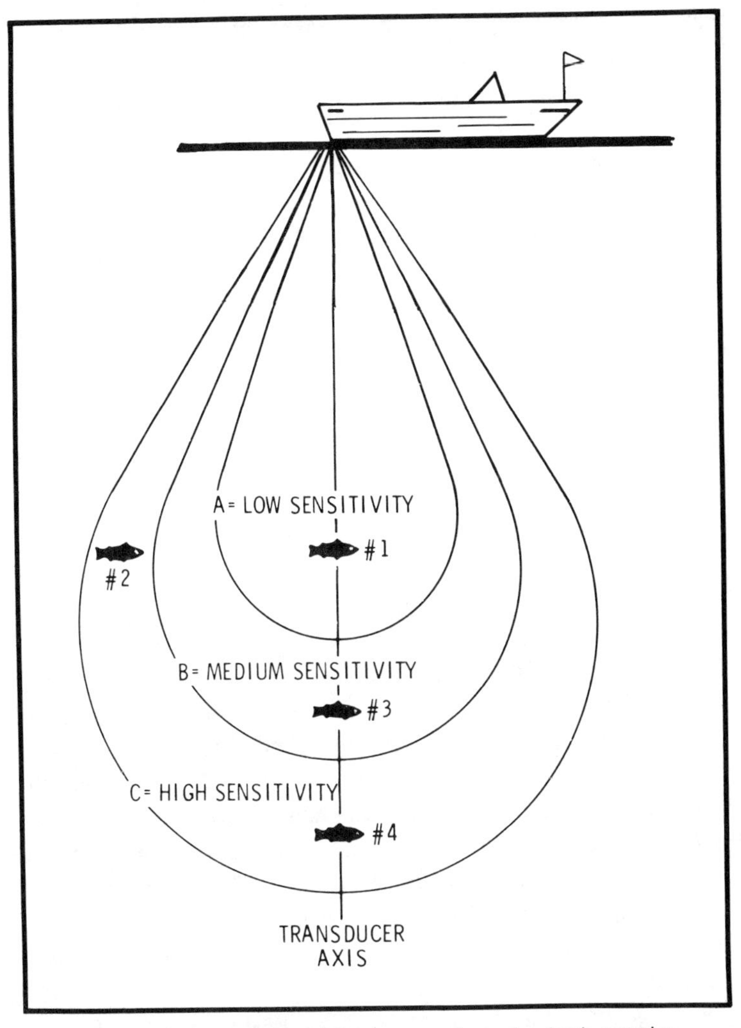

Fig. 6-3. A cross section of the beam pattern of a depth sounder.

Printing Office at the following address: Superintendent of Documents, Government Printing Office, Washington, D.C. 20402. You can, however, make your own chart for use in treasure hunting by using part of a county or highway map as a guide. One way to do this is to plot the position of each depth on a map as you circle the lake (Fig. 6-4). Check the positions of each depth by sighting toward two points on land in two directions. Check all the locations that might be of interest to you, such as good fishing areas and underwater objects.

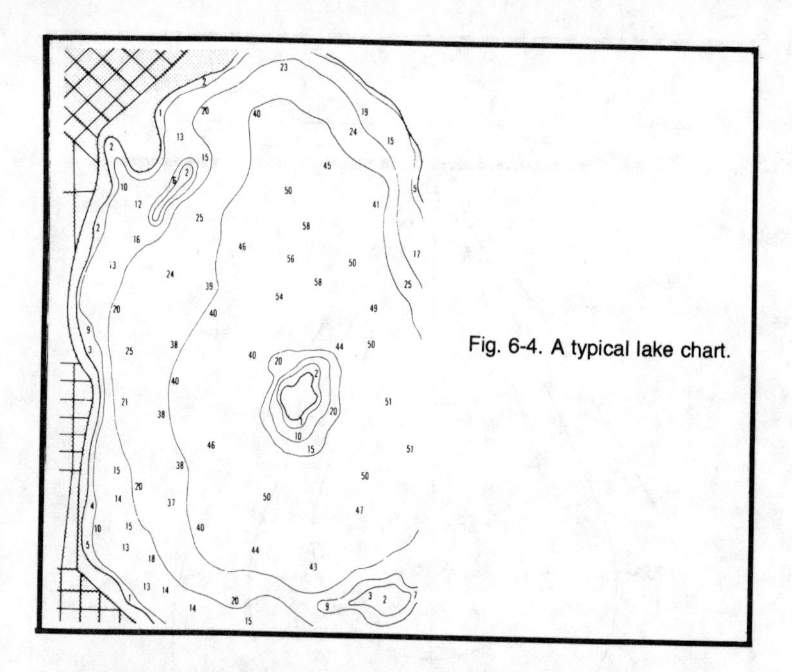

Fig. 6-4. A typical lake chart.

UNDERWATER METAL DETECTORS

Unlike land treasure hunting, most underwater treasure hunting limits the amount of time one can spend working on a treasure site. Weather conditions dictate when a diver can safely search. Also, underwater treasures are constantly moving, being covered and uncovered. A heavy storm can move tons of sand or mud on the ocean's bottom in only a matter of minutes. Therefore, any device that can shorten searching time will be highly beneficial to the treasure hunter. The underwater metal detector is one instrument that can accomplish this.

In general, underwater metal detectors are less expensive and easier to operate than most other gear associated with underwater treasure hunting. Depth sounders, radar, and magnetometers can run into thousands of dollars, but a good underwater detector costs less than $500.

Shipwreck sites are reasonably easy to locate because records are obtainable for many wrecks, but most of the old wooden vessels have probably long since disintegrated, leaving no visible evidence of their existence. That's where an underwater metal detector will come in—to penetrate sand, mud, wood, or coral and disclose the location of metal from an old wreck.

There are two kinds of underwater metal detectors: beat frequency oscillator types or transmitter-receiver types. Most have remote heads with a connecting cable. The search head may be lowered into waters by a diver to depths up to the length of the cable. However, another person will have to be stationed *above* the water to monitor the find indicator.

In shallow water, a regular land metal detector may be used because the search heads on most brands are waterproof.

Chapter 7
Finding Treasure-Hunting Waters

The information available on treasure-hunting waters is plentiful. Usually, it's just a matter of tracking down the data you need.

GOLD-BEARING WATERS

It has been reported that half the refined gold of the entire world lies scattered on the bottoms of lakes, rivers, and oceans. Probably most of the gold ore also lies somewhere near, or in, waters.

When someone thinks in terms of prospecting for gold his mind very likely turns to thoughts of the 49ers during the California Gold Rush or possibly to some remote stream in Alaska. Actually, some of the richest gold deposits ever found were located in the East—the James and Rappahannock River valleys of Virginia and a small region in the extreme northwestern part of Georgia. Most of the mining went on over 100 years ago but some commercial gold mining was still going on in the early days of this century. A modern day treasure hunter with a dredging machine and a gold pan may still be able to find enough gold around these areas to make the practice worthwhile.

Here's a list of a few of the most likely places to find gold:

Alabama—Two creeks in Alabama are known gold producers: Chulafinnee Creek in Cleburne County and Mulberry Creek in Chilton County.

Arizona—The Colorado River, cutting across the northwest corner of the state, is a good place to begin.

California—One of the richest gold producing states since the 49ers. Some of the better rivers and streams include the Klamath, Pit, Trinity, Feather, Yuba, Bear, Sacramento, Cosumnes, Mokelumne, Calaveras, Stanislaus, Tuolumne, San Joaquin, Merced, Chowchilla, Fresno, Placerita, and San Gabriel.

Colorado—The South Platte River in the northern part of the state yields gold; the Arkansas River in the eastern part of the state is also good.

Georgia—Another rich gold-bearing area. Try the streams in Lumpkin County which include Baggs Branch, Etowah Creek, Chestatee Creek, and Yahoolah Creek.

Idaho—Idaho has two good gold-bearing streams which should be of interest to the weekend prospector; they are the Snake and the Salmon Rivers.

Maine—The Swift river in Franklin County has yielded some gold in the past and should still provide action for weekend prospectors.

Nebraska—The gold-bearing Platte River, which runs the entire length of the state, is a good possibility.

Nevada—There are several good rivers in this state known to produce gold. Try the Walker, Carson, Humboldt, and Reese Rivers.

New Hampshire—Try sections of Indian Creek in Coos County.

New Mexico—The San Juan River, looping into the northern section of the state, is a good gold producer, as is the Rio Grande running north and south.

North Carolina—The first gold discovered in America was probably found in North Carolina. The Valley River in Cherokee County is a good spot to begin looking.

Oregon—Try the Calapooya and Rogue rivers; both are located in the western part of the state.

Pennsylvania—Any of the streams near the southern state line should be possibilities—especially those in the extreme southwest.

South Carolina—Just south of the border of North Carolina lie two gold-bearing streams: Whitewater and the Texaway. Both are located in Oconee County.

South Dakota—The Cheyenne River, which runs nearly across the entire state, is a gold-bearing river; try sections in the southwest corner of the state.

Tennessee—Two streams, both located in Monroe County, are good possibilities: Coker Creek and Whippoorwill.

OTHER TREASURE-BEARING WATERS

If you want to dive for sunken treasures, this country has sunken ships off both coasts as well as in the Great Lakes and other bodies of water. A reference service report of available shipwreck data can be obtained from the General Services Administration, Archives and Records Service, Washington, D.C. 20408. "Sources of Information Regarding Wrecks in the Great Lakes" is another bulletin available that can help the treasure hunter locate sunken ships in the Great Lakes. This bulletin is available from the U.S. Army Engineer District, Lake Survey, Corps of Engineers, 630 Federal Building, Detroit, Michigan 48226.

You can also get lots of treasure-hunting data from the Superintendent of Documents, U.S. Government Printing Office, Washington, D.C. 20402. There is a small charge for these publications but they are well worth it. One bulletin that will be of help to treasure hunters is "A Descriptive List of Treasure Maps and Charts in the Library of Congress" (No. 64-60033). Maps showing explorers' routes, trails, and early roads in the United States may not be of too much help in finding treasure-bearing waters, but no treasure hunter should be without them. A book containing such information has been compiled by Richard S. Ladd: *Hunting For Fact* (No. 62-60066; U.S. Government Publication No. GP 3.22:F11/963).

Authentic maps, showing the approximate location of sunken ships, are readily available from various sources. A typical treasure map is shown in Fig. 7-1. This is a map of three of the Great Lakes showing the location of over a dozen shipwrecks. On Lake Superior, for example, the *H.B. Smith* was lost in 1913 (approximate location indicated by block 1 in Fig. 7-1). The *Leafield* sank the same year in the vicinity of block 2. The *LaRonde* sank at block 3 in 1763. The *Algona* and the *Sunbeam* were lost at blocks 5 and 6, respectively, sometime during the 1880s. The remains of these vessels will continue to provide treasure finds for many years to come.

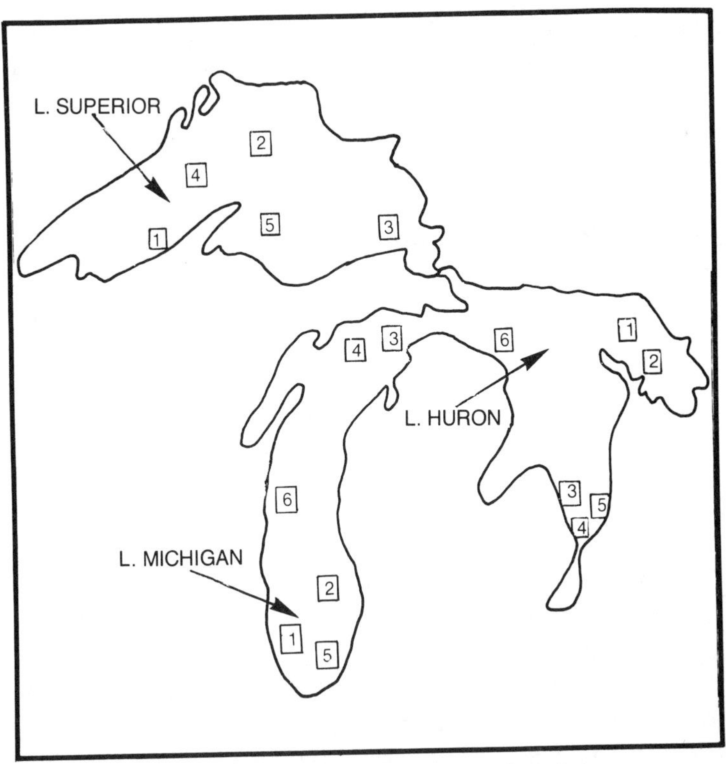

Fig. 7-1. Shipwreck sites on three of the Great Lakes.

In Lake Michigan the *Vernon* sank in 1887 at block 6 and the *Hercules* sank in 1918 at block 5. Later years (as late as 1953) brought about other disasters on the Lake, indicated by the remaining blocks.

Lake Huron also had her share of disasters. The first ship to be lost was the *Griffin*, lost in 1679 at block 1. Then the *Asia* was lost in 1882 (block 2), and the *Bruno* (block 6) went down in 1890. The *Price* (block 3), the *Argus* (block 4), and the *I. Carruthers* (block 5) were all lost in 1913.

These are just a few of the wrecks in the Great Lakes where millions in shipwreck treasure lie. But before treasure hunting in these waters, don't forget to check into the various regulations—like salvage rights—before undertaking your search.

Don't overlook local residences around waterfronts for information leading to a valuable treasure find. Many of these tales are going to be a little tall, but a little extra research on your own just might turn up that million dollar find. Ask intelligent questions and then check out any likely leads—at the library first and then in the water.

There are several books available (see Appendix 3) which give leads to underwater treasures. One such book is *Gold Locations of the United States* by Jack Black. It lists more than 1000 known gold locations. Another book, *Directory of Buried or Sunken Treasures and Lost Mines of the United States*, lists over 2269 lost treasures worth over $3,675,000. It lists the type of treasure lost, the approximate dollar value, and the approximate location.

Historical books can also give you treasure-hunting leads. Many treasure hunters overlook these sources. If you can read between the lines, you may be able to discover some very rich treasure-hunting waters.

For example, a treasure hunter from Atlanta, Georgia, had done some research on Civil War battles and had discovered that Fort Branch, on the banks of the Roanoke River near Hamilton, North Carolina, is an overlook and was chosen by the Confederacy because of its position. He suspected that the river itself—some 100 feet down from the fort—contained a lot of shells and guns. Three divers and the treasure hunter set out for the area and after just 20 minutes of diving they found numerous armament shells on the bottom of the river. Further investigation revealed some sort of cart—perhaps an ammunition cart—on the bottom also. The first day's diving uncovered over 150 pieces of ammunition and 12 different types of shells, but the divers wanted to return at a later date for more exploring.

The next outing turned up several cannons, some of which were in very good condition and were valued at about $100,000. Three of the cannons were placed on display to raise money to restore the fort.

Though some treasures are found by chance, most are found by careful research so anyone seriously interested in treasure hunting must do his homework before taking to the water.

Chapter 8
Techniques For Improving Your Luck

After you have gained some experience at treasure hunting, chances are you're going to run across some techniques that will increase your luck. Perhaps you have learned that most gold dust will collect on the outside bend in a stream or creek so you don't waste time searching the inside bend. Maybe you have decided that most treasures found around old homes will more than likely be found embedded in stone foundations, fireplaces, or similar locations where a fire would not damage them. Or maybe you are using a technique you have heard about—or read about. But it really doesn't matter how you learn a technique. The important thing is that it gets results.

SOME COMMONSENSE METHODS

A lot of techniques are just a matter of commonsense, a matter of thinking about likely treasure sites.

For example, when burying money was popular some decades ago, one of the favorite spots was at the bottom of a fence post. Usually a fence post was removed, the hole was dug a little deeper to accomodate the can or container holding the treasure, and then the post was placed back into position—over the container—and finally the earth around the post was tamped to secure it in place. So when searching around any old homesteads always check around the fence posts with your metal detector. In many instances, the fence posts may have rotted, leaving little evidence of the original location. However, there is a good chance you'll run across some fence wire with your metal detector; just follow this fence line with the search head. If there is any treasure at any of the original post holes, your detector should turn it up. A detector with a discriminator circuit should be used in a case like this: The signals from the fence wire will probably mask signals from the treasure when using a regular detector circuit.

One favorite trick of those who buried money was to drive a nail in the underside of a tree limb, attach a string to the nail, and place a weight on the other end of the string (like a plumb bob). They would then bury their treasure at the point where the weight touched the ground. Anytime they wanted to recover the cache they would locate the nail in the tree limb, tie a string with a weight attached, and dig where the weight indicated. Any treasure hunter who suspects such a treasure should keep an eye out for nails on the underside of limbs; checking limbs out with the metal detector is also a worthwhile practice.

You should also consider that whoever hid the treasure in the first place had to have some landmarks to guide him back to the hiding place. Look for large unusual rocks or a large flat rock with a marking carved on it. In Powell's Fort Valley, located in Northern Virginia, one Mr. Powell (for whom the fort was named) supposedly buried a cache of silver under a large flat rock back in the 1700s and carved a horseshoe-shaped design on the rock. Some 200 years later a 12-year-old girl was in the vicinity picking huckleberries and came upon the rock but didn't know the story. She told of the rock at dinner that same evening. Her father had heard of the legend and immediately set out in search of the rock the following morning, but it was never found again. Perhaps someone with a metal detector could locate this find and turn legend into fact!

When searching around old homesteads, try to locate the site of the chicken house. Years ago, chickens served as excellent burglar alarms since they would raise quite a racket when disturbed at night. Therefore, many people buried their money under a floorboard in the chicken house.

Many times treasures were buried halfway between two points. Put yourself in the shoes of the person who originally hid the treasure. What landmarks would they have used?

Obviously, a lot of commonsense techniques depend upon prior research. In fact, about 90% of all major treasures are located because of careful research *before* taking to the field.

USING DOWSING INSTRUMENTS

One of the oldest known methods of searching for buried treasures or valuable mineral deposits is that of dowsing. The early Spanish explorers used dowsing instruments in their extensive mining operations throughout the southwestern part of this country and Mexico. They had no modern metal detectors, but they always seemed to discover the richest and most productive areas.

Marco Polo recorded seeing dowsing instruments during his travels to China in the 13th century. Many other references to the use of dowsing instruments for locating metals and minerals are found in Germanic, English, and French writings of the early Middle Ages. And our troops in Viet Nam reported that some of the peasants used crude dowsing rods to find land mines and ammunition dumps.

Obviously, a dowsing instrument can save the treasure hunter a lot of time and wasted effort. It can sense the presence of gold or silver from a great distance and is able to indicate their general direction. One such dowsing instrument is shown in Fig. 8-1. Dowsing instruments are not made to take the place of electronic metal detectors; rather they are a tool by which the treasure hunter can increase his percentage of recoveries far beyond that of any other known method.

Unfortunately, dowsing instruments can be obtained from only a few sources. One such source is Carl Anderson, Box 13441,

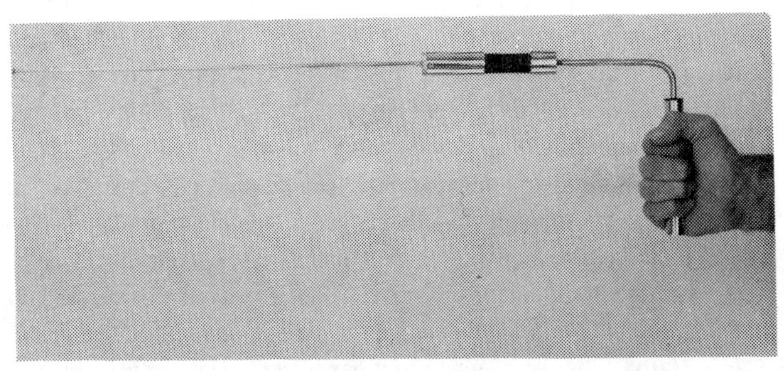

Fig. 8-1. Universal Antenna Rod dowsing instrument.

Tampa, Florida 33611. Carl has designed and built a line of dowsing instruments which are probably the most sensitive long-range locating instruments on the market today. His instruments have been used by amateurs and professionals all over the world—many of whom have reported great success.

The operation of one of Carl's dowsing instruments is really not as mysterious as it sounds. It is a well known fact that every substance gives off radiation, or rays, produced by the phenomenon of nuclear resonance. This radiation is of a different wavelength for each different material. Just as a radio picks up only radio waves from the particular station to which it is tuned, Carl Anderson's instruments pick up and react only to the radiations of certain metals and other designated materials. They are built to pick up only specific wavelengths of radiation, so they will react only to those certain substances.

His dowsing instruments draw their power directly from the electrostatic energy of your own body. The existence of this energy is well known and can be registered and measured by use of a simple galvanometer. When you grasp one of the instruments it is charged with your bodily electrostatic energy. The mineral crystal load in the instrument amplifies this energy and directs the powerful charge through the point or needle. The instrument, when so charged, will pick up and react to the radiation from gold or silver and will pull towards that direction.

Many people may then question if these instruments would be attracted by silverware in people's homes which may prevent it from pulling to a real treasure buried in the ground. Actually, when a treasure lies buried for several years the attractive radiations build up and concentrate in the soil immediately surrounding the treasure. This creates a halo around the treasure which is very much like an intense electromagnetic field around the treasure. The dowsing instrument will be much more strongly attracted to this concentrated buildup than to any unburied gold or silver. For this same reason, dowsing instruments will normally detect concentrations of buried treasure before it will those which are not buried.

The instruments will "search" through rock, concrete, water, dirt, and even iron. The distance these instruments will pick up treasures depends upon the size of the object, depth, and the skill of the operator. The top ranges are measured in miles for valuable ore deposits and large treasure caches. Smaller treasure will be picked up at distances which depend upon the size of the cache.

Users of dowsing instruments have even claimed to locate paper money, but at lesser distances than when searching for metals. Carl Anderson claims that his dowsing instruments can be attracted to a medium-sized cache of paper money at only about 10% to 25% of the distance from which you could pick up gold or silver.

These instruments are designed to work to only gold, silver, or paper money. They will not work to gas pockets, tree roots, damp ground, or tin foil. They will, however, locate gold or silver that may be buried in an iron pot. And these instruments are designed for use with large ore deposits. But if you are working an area where there are a lot of scattered coins, the coins usually can be found faster by using an electronic metal detector.

Dowsing instruments work better for some people than for others because some people have a low level of body electrostatic energy. But Carl claims that even those with a low level of electrostatic energy can get good results with practice—and that those who can't learn at all are rare. There is even a course on dowsing instruments held in Wilmington, North Carolina, that has reported over 100 students enrolling without one failure yet!

The Universal Antenna Rod dowsing instrument (Fig. 8-1) was designed to pick up even the faintest or most distant attractive rays. If you shorten the antenna, the instrument will react only to sources of attraction that lie closer by. The farther you extend the antenna, the greater the range. When the antenna is extended to its full length of 18 inches, the instrument can pull to large treasures or mineral deposits from many miles away. Smaller treasures will, of course, be found at lesser distances. You can even judge the distance of the treasure by shortening the antenna until it will no longer pick up the attraction.

Very good results may be obtained by using the 24-inch fiber handles. Just hold them as you would the old divining rods and turn slowly in a complete circle so as to face in different directions. When you face in the direction of a source of mineral or metal attraction, you will see and feel the instrument dip or pull lightly. If there is a second treasure within range, your instrument will again dip when you face that direction. Thus, you may ascertain the presence and directions of several treasures in the same area and choose which one you wish to follow.

Fig. 8-2. Operating the Antenna Rod dowsing instrument. It will swing to the right or left and will point in the direction of the metal or mineral for which the user is searching.

A Universal Antenna Rod has a magnetic tube charged with a special load which allows the instrument to be adjusted to pick up only certain wavelengths of attraction. When you install a small sample of the material you are searching for in the sample compartment at the tip of the tube, the instrument is then set to pick up only the attractive rays from the same kind of material. For example, if you are looking for gold, you must insert a small sample into the sample compartment. The instrument will then pull only to gold or gold ore. You can use this instrument to search for gold, silver, mercury, lead, copper, diamonds, paper money, oil, water, or any other substance. The rod is mounted on a bearing handle so it can swing freely, thereby reacting strongly to even the slightest attraction.

To operate the Universal Antenna Rod, hold it as shown in Fig. 8-2. It will swing to the right or left and will point in the direction of the metal or mineral you are searching for. When directly over the treasure, it will swing in a circle. The price for these instruments is about $250. They come complete with comprehensive operating instructions as well as samples of gold, silver, and other minerals.

The Spanish dip needle (Fig. 8-3) has been used by professional prospectors and treasure hunters for many years. It has proven itself in many areas of the world and was first used in the New World by the Spanish in the 16th and 17th centuries.

The Spanish dip needle has been known as the Mexican dip needle and also the Waterman. Many users report detecting large treasures or mineral deposits from several miles away. Smaller treasures can be traced from lesser distances, depending upon the size of the treasure.

The main body of the instrument is charged with a strong mineral load capable of multiplying the electrostatic energy of your body several times. Separate needles are provided for gold, silver, and paper money. When used with 24-inch fiber handles, the operator holds them tilted forward at an angle and then turns slowly so as to face every direction. When the operator faces a source of gold or silver attraction, the instrument will pull down or forward slightly. When directly over the source of attraction, it will indicate that spot by operating in a circular pattern.

When your dowsing instrument picks up an attraction and indicates the direction of the source of the attraction, make a line on

Fig. 8-3. The Spanish dip needle has been used by professional prospectors and treasure hunters for many years.

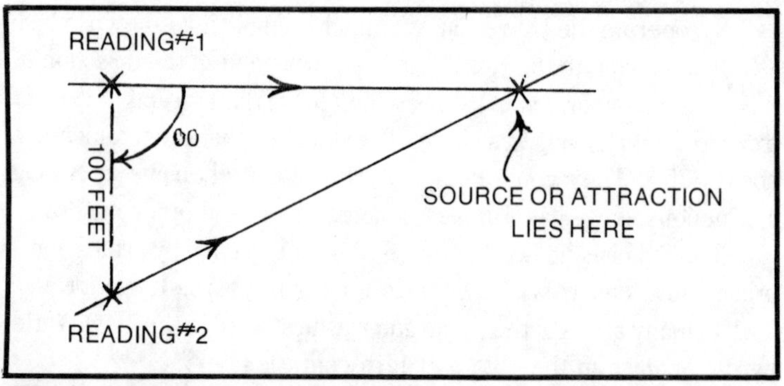

Fig. 8-4. A method of pinpointing a source of attraction using a dowsing instrument.

the ground to show that direction. Then move over about 100 feet at right angles to the direction indicated on the ground. Take another reading, again marking the direction on the ground. If there is an obvious difference in the direction of the two lines, the source is nearby—less than 100 yards. The source will be at the intersection of the two lines. You can estimate where this point is (Fig. 8-4).

If the source lies at a greater distance, the difference in the directions of the two readings will not be too noticeable. In such a case, take a good compass reading to note the direction of the attraction. Then move over at approximately a right angle to the direction of the attraction, a distance of several hundred yards or a mile or so. Take another compass reading on the direction of the attraction. Using a protractor, you can then plot the two directions on a map or aerial photo of the area, as shown in Fig. 8-5. The area where the two lines intersect is where the source of the attraction would be found.

To use the second method you will need to take an accurate reading from your instrument and then from your compass. Your calculations will be only as accurate as these readings.

USING ULTRAVIOLET LIGHT

Battery-powered ultraviolet lamps have aided prospectors, miners, and other treasure hunters in the location of ore deposits and other valuable treasures for many years. The use of ultraviolet lamps in detecting scheelite, for example, makes the location of it nearly as easy as locating one white rock lying on black sand. The method used is based on the fact that the fluorescent color of

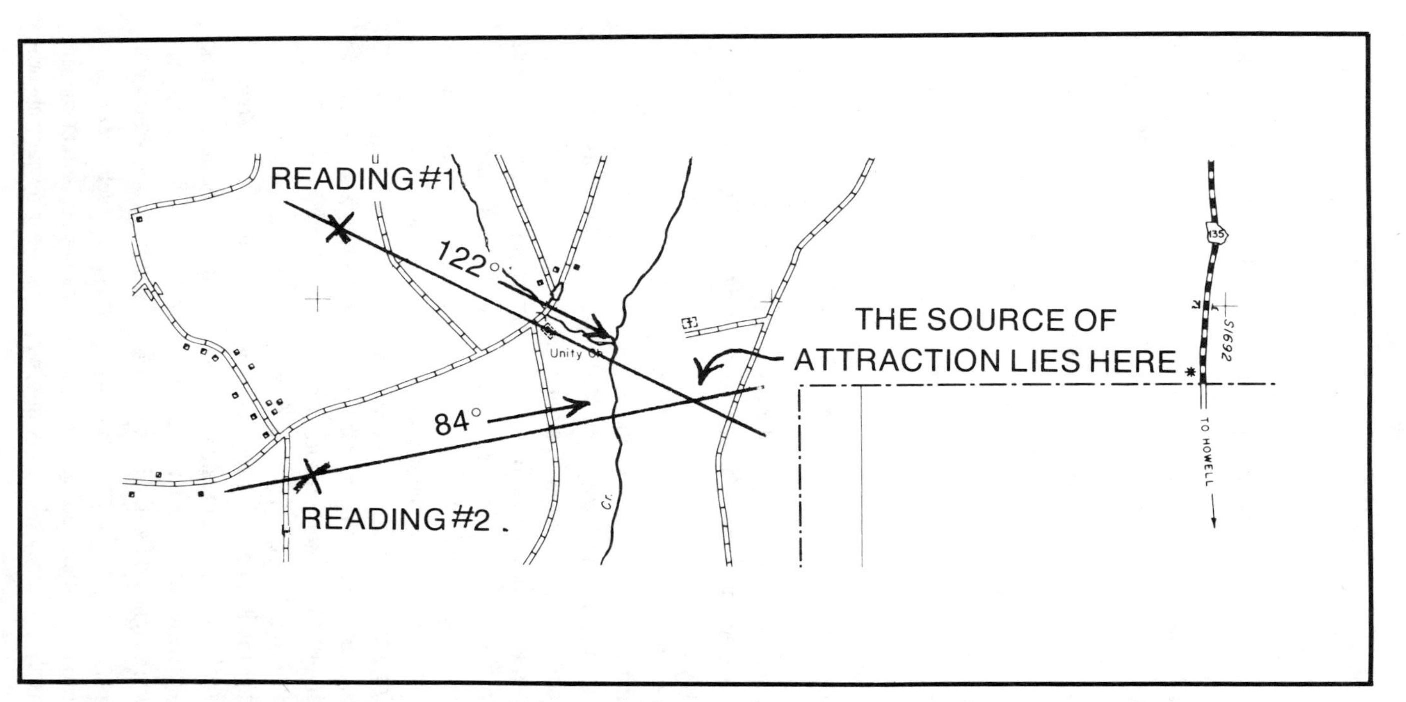

Fig. 8-5. Another method of pinpointing a source of attraction using a dowsing instrument and a compass.

scheelite is directly related to the amount of molybdenum contained in the crystal lattice.

Any good observer equipped with a good ultraviolet light source can determine the molybdenum content of scheelite. In general, specimens of scheelite which fluoresce a distinct blue contain less than 0.35% molybdenum; those which fluoresce white contain roughly 1.0%; and those which fluoresce yellow contain more than 1.0% molybdenum.

To further help the users of ultraviolet lamps a "Scheelite Fluorescence Analyzer Card" is available from Ultra-viotet Products, Inc., San Gabriel, California 91778, for about $15. This card provides an easy, accurate method of estimating the percentages of molybdenum in scheelite in accordance with the U.S. Bureau of Mines.

There are also many other uses for battery-powered ultraviolet lamps. For example, in recent years they have been used for prospecting and detecting uranium in various minerals. Prospectors now very frequently rely upon ultraviolet lamps rather than the usual chemical and physical techniques employed for uranium exploration and analysis.

A number of minerals containing uranium can be located by shining ultraviolet light on them; a yellowish-green fluorescence indicates uranium content. You will probably find that uranium ores are more easily traced with a portable ultraviolet lamp than with a conventional Geiger counter because the lamp provides a valuable aid in tracing veins of primary ore by the yellowish-green fluorescence of the secondary uranium minerals. The fluorescent uranium minerals are the uranium phosphates, arsenates, sulfates, and others which fluoresce with a strong yellow-green color.

An ultraviolet lamp can also be used to identify thorium and rare earth minerals. Thorium and rare earth minerals will set a Geiger counter to ticking at a very rapid pace. But they contain no uranium and are totally nonfluorescent. So when using your Geiger counter, take along an ultraviolet lamp to check the sample before beginning any extensive digging. Thorium and certain rare earth minerals are easily recognized by the dull to emerald green they show under an ultraviolet lamp with its filter removed. This green color seen without the filter is not fluorescence but rather an optical effect caused, primarily, by neodymium, a cerium-group rare earth which absorbs yellow light, plus most of the blue and violet rays, and reflects the green.

Another use for the ultraviolet lamp is to identify mercury ores. Since the lamp uses mercury vapor to generate ultraviolet light, any mercury vapor given off by an ore sample will absorb this light and can thus be identified. Begin the testing process by setting up a specially coated screen—usually of brightly fluorescent willemite (crystalline zinc silicate) on a cardboard or wooden sheet. Place the ore sample between the screen and the ultraviolet lamp. Continue by heating a crushed sample of the ore while the lamp is turned on. While the mercury vapors themselves are invisible, they will absorb the ultraviolet light and cast a visible shadow on the fluorescent screen if any mercury is present.

WARNING

Mercury vapors are highly toxic. Do not inhale them.

Chapter 9
Determining The Value Of Your Finds

This chapter is devoted entirely to information regarding the value of items you may find. Knowing the worth of your finds will enable you to decide which treasures you really want to hunt.

COINS

Since coins are found everywhere it is not surprising that they are the number one target of treasure hunters. Coins are found every day by people with metal detectors and many of the coins are worth several times their face value.

Coin collectors generally categorize coins by condition. There are three conditions:

Proof condition: Proof coins are specially struck for coin collectors by the U.S. Mint out of flawless dies. All proof coins will possess a mirrorlike surface and have absolutely no imperfections.

Uncirculated condition: These coins are in new condition and have never been circulated. All will be of high quality and show a brilliance.

Circulated condition: The better or "nice" circulated coins will have a clearly defined date and all basic details will be clear. The nice circulated condition coin will be the condition of most coins found by people using metal detectors.

The following is a list of U.S. coins. The prices listed are the rates normally paid by collectors when the coins are in nice circulated condition. However, if you sell such coins to dealers you can expect to receive up to 50% less. Chapter 10 will tell you where and how to get the most cash return from your finds, but if you need cash quickly, the coin dealers are probably your best bet.

Indian Cents (1857-1909)

1857....$8.95
1858 LL....$8.95
1858 SL....$8.95
1859....$4.25
1860....$3.50
1861....$8.50
1862....$3.25
1863....$3.25
1864 C.N.....$5.95
1864 Bronze....$2.95
1864 "L"....$12.95
1865....$2.95
1866....$10.50
1867....$10.50
1868....$10.50
1869....$17.95
1870....$16.95
1871....$19.95
1872....$22.95
1873....$5.50
1874....$4.95
1875....$4.95
1876....$8.95
1877....$89.95
1878....$8.95
1879....$2.50
1880....$1.95
1881....$1.95
1882....$1.95
1883....$1.95
1884....$2.50
1885....$3.95
1886....$2.50
1887....$1.25
1888....$1.25
1889....$1.25
1890....$1.25
1891....$1.25
1892....$1.25
1893....$1.25
1894....$1.95
1895....$0.95
1896....$0.95
1897....$0.95
1898....$0.95
1899....$0.95
1900....$0.95
1901....$0.95
1902....$0.95
1903....$0.95
1904....$0.95
1905....$0.95
1906....$0.95
1907....$0.95
1908....$0.95
1908 S....$18.95
1909....$1.25
1909 S....$69.50

Lincoln Cents (1909-1976)

1909....$0.50
1909 VDB....$2.95
1909 S....$29.95
1909 S VDB....$139.95
1910....$0.35
1910 S....$6.95
1911....$0.35
1911 D....$3.50
1911 S....$9.95
1912....$0.40
1912 D....$2.95
1912 S....$8.95
1913....$0.40
1913 D....$1.95
1913 S....$5.95
1914....$0.40
1914 D....$45.00
1914 S....$8.95
1915....$0.95
1915 D....$0.75
1915 S....$6.95
1916....$0.35
1916 D....$0.50
1916 S....$0.95
1917....$0.35
1917 D....$0.50
1917 S....$0.50
1918....$0.35
1918 D....$0.50
1918 S....$0.50
1919....$0.35
1919 D....$0.45
1919 S....$0.40
1920....$0.35
1920 D....$0.50
1920 S....$0.50
1921....$0.40
1921 S....$0.95
1922 D....$5.95
1923....$0.35

1923 S	$1.95
1924	$0.35
1924 D	$9.95
1924 S	$0.95
1925	$0.35
1925 D	$0.50
1925 S	$0.50
1926	$0.35
1926 D	$0.50
1926 S	$3.95
1927	$0.35
1927 D	$0.50
1927 S	$0.75
1928	$0.35
1928 D	$0.40
1928 S	$0.75
1929	$0.35
1929 D	$0.40
1929 S	$0.40
1930	$0.30
1930 D	$0.35
1930 S	$0.45
1931	$0.50
1931 D	$2.95
1931 S	$29.95
1932	$1.95
1932 D	$0.75
1933	$0.70
1933 D	$2.50
1934	$0.25
1934 D	$0.35
1935	$0.25
1935 D	$0.35
1935 S	$0.35
1936	$0.25
1936 D	$0.35
1936 S	$0.35
1937	$0.25
1937 D	$0.35
1937 S	$0.35
1938	$0.25
1938 D	$0.50
1938 S	$0.50
1939	$0.35
1939 D	$0.50
1939 S	$0.35
1940	$0.20
1940 D	$0.20
1940 S	$0.20
1941	$0.20
1941 D	$0.20
1941 S	$0.20
1942	$0.20
1942 D	$0.20
1942 S	$0.20
1943	$0.35
1943 D	$0.50
1943 S	$0.50
1944	$0.20
1944 D	$0.20
1944 S	$0.20
1945	$0.20
1945 D	$0.20
1945 S	$0.20
1946	$0.20
1946 D	$0.20
1946 S	$0.20
1947	$0.20
1947 D	$0.20
1948	$0.20
1948 D	$0.20
1948 S	$0.20
1949	$0.20
1949 D	$0.20
1949 S	$0.20
1950	$0.20
1950 D	$0.20
1950 S	$0.20
1951	$0.20
1951 D	$0.20
1951 S	$0.20
1952	$0.20
1952 D	$0.20
1952 S	$0.20
1953	$0.20
1953 D	$0.20
1953 S	$0.20
1954	$0.35
1954 D	$0.20
1954 S	$0.20
1955	$0.20
1955 D	$0.20
1955 S	$0.50
1956	$0.20
1956 D	$0.20
1957	$0.20
1957 D	$0.20
1958	$0.20
1958 D	$0.20
1959	$0.20
1959 D	$0.20
1960	$0.20
1960 Small Date	$4.95
1960 D	$0.20
1960 D Small Date	$0.40
1961	$0.20

1961 D	$0.20	1970 S	$0.20
1962	$0.20	1970 Small Date	$2.95
1962 D	$0.20	1971 P	$0.20
1963	$0.20	1971 D	$0.20
1963 D	$0.20	1971 S	$0.20
1964	$0.20	1972	$0.20
1964 D	$0.20	1972 D	$0.20
1965	$0.20	1972 S	$0.20
1966	$0.20	1973	$0.20
1967	$0.20	1973 D	$0.20
1968	$0.20	1973 S	$0.20
1968 D	$0.20	1974	$0.20
1968 S	$0.20	1974 D	$0.20
1969	$0.20	1974 S	$0.20
1969 D	$0.20	1975 P	$0.20
1969 S	$0.20	1975 D	$0.20
1970	$0.20	1975 S Proof	$24.95
1970 D	$0.20	1976 S Proof	$19.95

Liberty Nickels (1883-1912)

1883 without cents	$2.50	1898	$1.75
1883 with cents	$5.95	1899	$1.75
1884	$5.95	1900	$0.75
1885	$79.95	1901	$0.75
1886	$29.95	1902	$0.75
1887	$3.95	1903	$0.75
1888	$5.50	1904	$0.75
1889	$3.75	1905	$0.75
1890	$4.75	1906	$0.75
1891	$3.75	1907	$0.75
1892	$3.95	1908	$0.75
1893	$3.50	1909	$0.75
1894	$5.50	1910	$0.75
1895	$2.75	1911	$0.75
1896	$3.25	1912	$0.75
1897	$1.75	1912 D	$1.75
		1912 S	$29.95

Buffalo Nickels (1913-1938)

1913 Type 1	$2.50	1916	$0.95
1913 D Type 1	$4.95	1916 D	$3.95
1913 S Type 1	$6.95	1916 S	$3.95
1913 Type 2	$2.95	1917	$0.95
1913 D Type 2	$27.50	1917 D	$3.95
1913 S Type 2	$37.50	1917 S	$3.95
1914	$2.95	1918	$0.95
1914 D	$24.50	1918 D	$5.95
1914 S	$4.95	1918 S	$4.95
1915	$1.95	1919	$0.95
1915 D	$6.50	1919 D	$3.95
1915 S	$7.95	1919 S	$2.95

Date	Price
1920	$0.95
1920D	$3.50
1920 S	$1.95
1921	$1.50
1921 S	$9.95
1923	$0.75
1923 S	$1.95
1924	$0.75
1924 D	$2.95
1924 S	$4.95
1925	$0.95
1925 D	$3.95
1925 S	$2.95
1926	$0.75
1926 D	$2.50
1926 S	$6.95
1927	$0.75
1927 D	$1.50
1927 S	$1.50
1928	$0.75
1928 D	$0.85
1928 S	$0.95
1929	$0.50
1929 D	$0.75
1929 S	$0.75
1930	$0.50
1930 S	$0.75
1931 S	$3.95
1934	$0.35
1934 D	$0.50
1935	$0.35
1935 D	$0.50
1935 S	$0.50
1936	$0.35
1936 D	$0.50
1936 S	$0.50
1937	$0.35
1937 D	$0.35
1937 S	$0.50
1938 D	$0.50

Jefferson Nickels (1938-1976)

Date	Price
1938	$0.25
1938 D	$1.50
1938 S	$2.95
1939	$0.25
1939 D	$4.95
1939 S	$0.75
1940	$0.25
1940 D	$0.25
1940 S	$0.25
1941	$0.25
1941 D	$0.25
1941 S	$0.25
1942	$0.25
1942 D	$0.25
1942 Silver	$0.50
1942 S	$0.50
1943	$0.50
1943 D	$1.75
1943 S	$0.50
1944	$0.50
1944 D	$0.50
1944 S	$0.50
1945	$0.50
1945 D	$0.50
1945 S	$0.50
1946	$0.25
1946 D	$0.25
1946 S	$0.25
1947	$0.25
1947 D	$0.25
1947 S	$0.25
1948	$0.25
1948 D	$0.25
1948 S	$0.25
1949	$0.25
1949 D	$0.25
1949 S	$0.25
1950	$0.50
1950 D	$9.95
1951	$0.25
1951 D	$0.25
1951 S	$0.50
1952	$0.25
1952 D	$0.25
1952 S	$0.25
1953	$0.25
1953 D	$0.25
1953 S	$0.25
1954	$0.25
1954 D	$0.25
1954 S	$0.25
1955	$0.50
1955 D	$0.25
1956	$0.25
1956 D	$0.25
1957	$0.25
1957 D	$0.25
1958	$0.25
1958 D	$0.25
1959	$0.25

1959 D	$0.25
1960	$0.25
1960 D	$0.25
1961	$0.25
1961 D	$0.25
1962	$0.25
1962 D	$0.25
1963	$0.25
1963 D	$0.25
1964	$0.25
1964 D	$0.25
1965	$0.25
1966	$0.25
1967	$0.25
1968 D	$0.25
1968 S	$0.25
1969 D	$0.25
1969 S	$0.25
1970 D	$0.25
1970 S	$0.25
1971	$0.25
1971 D	$0.25
1971 S Proof	$3.50
1972	$0.25
1972 D	$0.25
1972 S Proof	$3.50
1973	$0.25
1973 D	$0.25
1973 S Proof	$3.50
1974 P	$0.25
1974 D	$0.25
1974 S Proof	$5.95
1975 P	$0.25
1975 D	$0.25
1975 S Proof	$4.95
1976 S Proof	$5.95

Mercury Dimes (1916-1945)

1916	$1.25
1916 D	$95.00
1916 S	$2.50
1917	$0.95
1917 D	$2.50
1917 S	$1.25
1918	$0.95
1918 D	$1.25
1918 S	$1.25
1919	$0.95
1919 D	$1.30
1919 S	$1.50
1920	$0.95
1920 D	$1.25
1920 S	$1.25
1921	$12.95
1921 D	$21.50
1923	$0.95
1923 S	$1.25
1924	$0.95
1924 D	$1.25
1924 S	$1.25
1925	$0.95
1925 D	$2.50
1925 S	$1.25
1926	$0.95
1926 D	$1.25
1926 S	$6.95
1927	$0.95
1927 D	$1.50
1927 S	$1.25
1928	$0.95
1928 D	$1.30
1928 S	$1.25
1929	$0.95
1929 D	$1.25
1929 S	$1.25
1930	$1.25
1930 S	$2.95
1931	$1.50
1931 D	$6.95
1931 S	$2.95
1934	$0.75
1934 D	$0.75
1935	$0.75
1935 D	$0.75
1935 S	$0.75
1936	$0.75
1936 D	$0.75
1936 S	$0.75
1937	$0.75
1937 D	$0.75
1937 S	$0.75
1938	$0.75
1938 D	$1.25
1938 S	$0.75
1939	$0.75
1939 D	$0.75
1939 S	$0.75
1940	$0.75
1940 D	$0.75
1940 S	$0.75
1941	$0.75
1941 D	$0.75
1941 S	$0.75
1942	$0.75

1942 D $0.75
1942 S $0.75
1943 .... $0.75
1943 D $0.75
1943 S $0.75
1944 .... $0.75
1944 D $0.75
1944 S $0.75
1945 .... $0.75
1945 D $0.75
1945 S $0.75
1945 Micro "S" $0.75

Roosevelt Dimes (1946-1976)

1946 .... $0.75
1946 D $0.75
1946 S $0.75
1947 .... $0.75
1947 D $0.75
1947 S $0.75
1948 .... $0.75
1948 D $0.75
1948 S $0.75
1949 .... $0.75
1949 D $0.75
1949 S $0.95
1950 .... $0.75
1950 D $0.75
1950 S $0.75
1951 .... $0.75
1951 D $0.75
1951 S $0.75
1952 .... $0.75
1952 D $0.75
1952 S $0.75
1953 .... $0.75
1953 D $0.75
1953 S $0.75
1954 .... $0.75
1954 D $0.75
1954 S $0.75
1955 .... $1.25
1955 D $0.95
1955 S $0.75
1956 .... $0.75
1956 D $0.75
1957 .... $0.75
1957 D $0.75
1958 .... $0.75
1958 D $0.75
1959 .... $0.75
1959 D $0.75
1960 .... $0.75
1960 D $0.75
1961 .... $0.75
1961 D $0.75
1962 .... $0.75
1962 D $0.75
1963 .... $0.75
1963 D $0.75
1964 .... $0.75
1964 D $0.75
1965 .... $0.25
1966 .... $0.25
1967 .... $0.25
1968 .... $0.25
1968 D $0.25
1968 S Proof $1.95
1969 .... $0.25
1969 D $0.25
1969 S Proof $1.95
1970 .... $0.25
1970 D $0.25
1970 S Proof $1.95
1971 .... $0.25
1971 D $0.25
1971 S Proof $1.95
1972 .... $0.25
1972 D $0.25
1972 S Proof $1.95
1973 .... $0.25
1973 D $0.25
1973 S Proof $1.95
1974 P $0.25
1974 D $0.25
1974 S $1.95
1975 P $0.25
1975 D $0.25
1975 S Proof $3.95
1976 S Proof $3.95

Washington Quarters (1932-1976)

1932 .... $2.25
1932 D $37.50
1932 S $37.50
1934 .... $1.35
1934 D $2.95
1935 .... $1.35
1935 D $1.35
1935 S $1.35
1936 .... $1.35
1936 D $1.35

1936 S	$1.35
1937	$1.35
1937 D	$1.35
1937 S	$4.95
1938	$1.35
1938 S	$2.25
1939	$1.35
1939 D	$1.35
1939 S	$2.25
1940	$1.35
1940 D	$2.25
1940 S	$1.35
1941	$1.35
1941 D	$1.35
1941 S	$1.35
1942	$1.35
1942 D	$1.35
1942 S	$1.35
1943	$1.35
1943 D	$1.35
1943 S	$1.35
1944	$1.35
1944 D	$1.35
1944 S	$1.35
1945	$1.35
1945 D	$1.35
1945 S	$1.35
1946	$1.35
1946 D	$1.35
1946 S	$1.75
1947	$1.35
1947 D	$1.35
1947 S	$1.35
1948	$1.35
1948 D	$1.35
1948 S	$1.35
1949	$1.35
1949 D	$1.35
1950 P	$1.35
1950 D	$1.35
1950 S	$1.35
1951	$1.35
1951 D	$1.35
1951 S	$1.35
1952	$1.35
1952 D	$1.35
1952 S	$1.35
1953	$1.35
1953 D	$1.35
1953 S	$1.35
1954	$1.35
1954 D	$1.35
1954 S	$1.35
1955	$1.35
1955 D	$1.35
1956	$1.35
1956 D	$1.35
1957	$1.35
1957 D	$1.35
1958	$1.35
1958 D	$1.35
1959	$1.35
1959 D	$1.35
1960	$1.35
1960 D	$1.35
1961	$1.35
1961 D	$1.35
1962	$1.35
1962 D	$1.35
1963	$1.35
1963 D	$1.35
1964	$1.35
1964 D	$1.35
1965	$0.75
1966	$0.75
1967	$0.75
1968	$0.75
1968 D	$0.75
1968 S Proof	$1.35
1969	$0.75
1969 D	$0.75
1969 S Proof	$1.35
1970	$0.75
1970 D	$0.75
1970 S Proof	$1.35
1971 P	$0.75
1971 D	$0.75
1971 S Proof	$1.35
1972 P	$0.75
1972 D	$0.75
1972 S Proof	$1.35
1973 P	$0.75
1973 D	$0.75
1973 S Proof	$1.35
1974 P	$0.75
1974 D	$0.75
1974 S Proof	$2.75
1976 P	$0.95
1976 D	$0.95
1976 S Proof	$3.95
1976 S40% Proof	$4.95
1976 S40% B.U.	$2.95

Walking Liberty Halves (1916-1947)

Date	Price	Date	Price
1916	$12.95	1936	$2.95
1916 D	$8.95	1936 D	$2.95
1916 S	$19.95	1936 S	$2.95
1917	$2.95	1937	$2.95
1917 D Obv	$6.95	1937 D	$3.50
1917 D Rev	$3.50	1937 S	$2.95
1917 S Obv	$7.50	1938	$2.95
1917 S Rev	$2.95	1938 D	$24.95
1918	$2.95	1939	$2.95
1918 D	$2.95	1939 D	$2.95
1918 S	$2.95	1939 S	$2.95
1919	$5.95	1940	$2.95
1919 D	$4.95	1940 S	$2.95
1919 S	$4.95	1941	$2.95
1920	$2.95	1941 D	$2.95
1920 D	$3.95	1941 S	$2.95
1920 S	$2.95	1942	$2.95
1921	$34.50	1942 D	$2.95
1921 D	$59.50	1942 S	$2.95
1921 S	$8.95	1943	$2.95
1923 S	$3.75	1943 D	$2.95
1927 S	$3.75	1943 S	$2.95
1928 S	$3.75	1944	$2.95
1929 D	$3.95	1944 D	$2.95
1929 S	$3.75	1944 S	$2.95
1933 S	$3.50	1945	$2.95
1934	$2.95	1945 D	$2.95
1934 D	$2.95	1945 S	$2.95
1934 S	$2.95	1946	$2.95
1935	$2.95	1946 D	$3.25
1935 D	$2.95	1946 S	$2.95
1935 S	$2.95	1947	$2.95
		1947 D	$2.95

Franklin Halves (1948-1963)

Date	Price	Date	Price
1948	$3.95	1954 D	$2.50
1948 D	$2.50	1054 S	$2.50
1949	$2.50	1955	$8.95
1949 D	$2.50	1956	$2.50
1949 S	$2.50	1957	$2.50
1950	$2.50	1957 D	$2.50
1950 D	$2.50	1958	$2.50
1951	$2.50	1958 D	$2.50
1951 D	$2.50	1959	$2.50
1951 S	$2.50	1959 D	$2.50
1952	$2.50	1960	$2.50
1952 D	$2.50	1960 D	$2.50
1952 S	$2.50	1961	$2.50
1953	$3.95	1961 D	$2.50
1953 D	$2.50	1962	$2.50
1953 S	$2.50	1962 D	$2.50
1954	$2.50	1963	$2.50
		1963 D	$2.50

Kennedy Halves (1964-1976)

Date	Price
1964	$2.50
1964 D	$2.50
1965	$1.50
1966	$1.50
1967	$1.50
1968 D	$1.50
1968 S Proof	$4.95
1969 D	$1.50
1969 S Proof	$4.95
1970 D Rare	$22.95
1970 S Proof	$9.95
1971	$0.95
1971 D	$0.95
1971 S Proof	$3.95
1972	$0.95
1972 D	$0.95
1972 S Proof	$3.95
1973	$0.95
1973 D	$0.95
1973 S Proof	$7.50
1974 P	$0.95
1974 D	$0.95
1974 S Proof	$6.95
1976	$0.95
1976 D	$0.95
1976 S40% Proof	$7.95
1976 S40% B.U.	$4.95
1976 S Clad Proof	$5.95

Morgan Dollars (1878-1921)

Date	Price
1878 7 Tail	$5.95
1878 8 Tail	$8.95
1878 CC	$9.95
1878 S	$5.95
1879	$5.95
1879 CC	$24.95
1879 O	$8.95
1879 S	$5.95
1880	$5.95
1880 CC	$29.95
1880 O	$5.95
1880 S	$5.95
1881	$5.95
1881CC	$59.95
1881 O	$5.95
1881 S	$5.95
1882	$5.95
1882 CC	$16.95
1882 O	$5.95
1882 S	$5.95
1883	$5.95
1883 CC	$16.95
1883 O	$5.95
1883 S	$7.95
1884	$5.95
1884 CC	$29.95
1884 O	$5.95
1884 S	$7.95
1885	$5.95
1885 CC	$89.95
1885 O	$5.95
1885 S	$7.95
1886	$5.95
1886 O	$5.95
1886 S	$19.95
1887	$5.95
1887 O	$5.95
1887 S	$9.95
1888	$5.95
1888 O	$5.95
1888 S	$24.95
1889	$5.95
1889 CC	$79.95
1889 O	$5.95
1889 S	$24.95
1890	$5.95
1890 CC	$12.95
1890 O	$5.95
1890 S	$5.95
1891	$5.95
1891 CC	$12.95
1891 O	$5.95
1891 S	$5.95
1892	$7.95
1892 CC	$24.95
1892 O	$7.95
1892 S	$9.95
1893	$24.95
1893 CC	$28.95
1893 O	$25.95
1893 S	$249.50
1894	$99.50
1894 O	$7.95
1894 S	$11.95
1895 O	$24.95
1895 S	$34.50
1896	$5.95
1896 O	$5.95

1896 S	$9.95	1901 O	$5.95
1897	$5.95	1901 S	$9.95
1897 O	$5.95	1902	$7.95
1897 S	$5.95	1902 O	$9.95
1898	$5.95	1902 S	$29.95
1898 O	$9.95	1903	$8.95
1898 S	$9.95	1903 O	$49.95
1899	$18.95	1903 S	$9.95
1899 O	$5.95	1904	$9.95
1899 S	$9.95	1904 O	$9.95
1900	$5.95	1904 S	$9.95
1900 O	$5.95	1921	$5.95
1900 S	$9.95	1921 D	$5.95
1901	$14.95	1921 S	$5.95

Peace Dollars (1921-1935)

1921	$19.95	1926 D	$7.95
1922	$5.95	1926 S	$5.95
1922 D	$5.95	1927	$19.95
1922 S	$5.95	1927 D	$12.95
1923	$5.95	1927 S	$12.95
1923 D	$5.95	1928	$99.50
1923 S	$5.95	1928 S	$7.95
1924	$5.95	1933	$19.95
1924 S	$8.95	1934 D	$7.95
1925	$5.95	1934 S	$8.95
1925 S	$8.95	1935	$10.95
1926	$8.95	1935 S	$8.95

Eisenhower Dollars (1971-1976)

1971 Clad B.U.	$1.95	1973 S Clad Proof	$4.95
1971 D Clad B.U.	$1.95	1973 S 40% Silver Proof	$29.95
1971 40% Silver B.U.	$4.95	1974 Clad B.U.	$1.95
1971 S 40% Silver Proof	$8.95	1974 D Clad B.U.	$1.95
1972 Clad B.U.	$1.95	1974 S Clad Proof	$4.95
1972 D Clad B.U.	$1.95	1974 S 40% Silver Proof	$9.95
1972 S 40% Silver B.U	$4.95	1974 S 40% Silver B.U	$6.50
1972 S 40% Silver Proof	$9.95	1976 P	$1.95
1973 Clad B.U.	$9.95	1976 D	$1.95
1973 D Clad B.U.	$9.95	1976 S Clad Proof	$4.95
1973 S 40% Silver B.U	$6.50	1976 S 40% Silver Proof	$9.95
		1976 S 40% Silver B.U	$5.95

BOTTLES

It has been estimated that there are over one billion dollars worth of old bottles lying around old dumps throughout the United States. Chances are you have overlooked several valuable ones on your treasure-hunting expeditions. The following list will help you recognize some of the more valuable ones on future expeditions.

Bottles That Contained Bitters

C.W. Abbott & Co., Baltimore, amber corker 2 1/2 in., 8 1/4 in., 10 1/2 in. $4 – 7.

Angelica Bitter Tonic, J. Triner, Chicago, flask amber corker, $12 – 25.

Burdock Blond Bitters, Buffalo, N.Y., corker 8 1/4 in. $12 – 18.

Cassin's Grape Brandy Bitters, green corker 9 3/4 in., $250 – 500.

H.P. Herb Wild Cherry Bitters, Wild Cherry on all four shoulders.

Morning Bitters Inceptum 5869, Patented 5869, three-sided amber corker 12 1/2 in., $125 – 200.

National Bitters, Patent 1867, Ear of corn, amber corker 12 1/4 in., $150 – 200.

Old Home Bitters, Laughlin Bros & Co., Wheeling, W. Va., amber corker, $125 – 175.

Old Homestead Wild Cherry Bitters, tall cabin, amber corker, $125 – 175.

Old Sachem Bitters and Wigwam Tonic, barrel 10 ribs top and bottom, amber corker, $85 – 125.

Soda Bottles

W.H. Burt, San Francisco, green, 7 1/4 in. $12 – 15.

Ownen Casey Eagle Soda Works, cobalt 7 in., $20 – 30.

D.S. & Co., San Francisco, cobalt 7 in., $25 – 40.

Nichols Flushing, L.I. Union Glass Works, Phila., soda dark jade 7 1/4 in., $20 – 35.

Hathorn Spring, Saratoga, N.Y., amber 9 1/2 in., $15 – 35.

M.R., Sacramento, cobalt 7 in., $40 – 70.

Superior Mineral Water, Union Glass Works, Phil., cobalt 7 in., $12 – 20.

Henry Winkle, Sac City, Cal., aqua 7 in., $25 – 35.

Woodburn Bottling Works, Woodburn, Ore., aqua 8 1/2 in., $3 – 5.

Zarembo Mineral Springs Co., Seattle, Wash., blue 7 1/4 in. $3 – 4.

Whiskey and Gin Bottles

Avan Hoboken & Co., Rotterdam, olive corker 11 1/4 in., $30 – 40.
E.G. Booz, originals, $8 – 15.
Dr. C. Bouvier's Buchu Gin, clear corker 11 3/4 in., $4 – 7.
Catto's Whiskey, olive green corker 9 in., $6 – 7.
Cresent Brandy, green corker 11 1/2 in., $3 – 5.
I.W. Harper, amber corker 9 1/2 in., $20 – 25.
Paul Jones (no seal), round amber corker 3 1/2 in., $2 – 3.

Beer and Wine Bottles

Anheuser Busch Co., aqua 9 1/2 in., $3 – 5.
Arnas, amber 8 in., $2 – 4.
Dr. Formanekc's Bitter Wine, round amber corker 10 1/2 in., $6 – 9.

Additional information on the value of bottles may be found in *Bottle Collector's Handbook and Pricing Guide* by John T. Yount, published by Educator Books, Inc., Drawer 32, San Angelo, Texas 76901. This book contains descriptions and pricing information on hundreds of valuable bottles found throughout the United States. Other helpful books are listed in Appendix 3.

It is also recommended that you obtain the following magazines from your local newstand: *The Antique Trader, Collector's World, Collector's Den, Hobbies Magazine* and *The National Bottle Gazette.* They will keep you posted on price changes and the most wanted bottles. The ads in these magazines will also give valuable pricing information.

CIVIL WAR RELICS

The interest in Civil War relic hunting and collecting increases each year. The following pricing guide should help treasure hunters determine the value of many of their finds.

Prices in this list are average prices usually paid by collectors. If you sell your finds to a dealer you can expect to receive about 50% less.

Belt Buckles

Small oval "US" belt plate	$35 – 60
Large oval "US" belt plate	$18 – 28

Waist belt plate for infantry noncommissioned officer's sword $25–45
Officer's sword belt plate $35–55
Oval waist belt plate, VMM-Volunteer, (Maine) $120–195
Sword belt plate, New York Militia regulation, square $125–175
Square Ohio Militia regulation sword belt plate $335–370
Oval belt plate with Pennsylvania state seal $220–325
Two-piece sword belt plate, "US" $150–235
Oval "CS" belt plate of somewhat different dimensions $235–455
Oval common "CS" belt plate, stamped brass $300–495
Oval "CS" cast-brass waist belt plate $500–650
"Virginia style CSA" cast-brass waist belt plate, square $205–350
Small model "CSA" cast-brass waist belt plate $160–285
"CS" rectangular common waist belt plate, brass $270–400
Two-piece "CS" common sword or saber belt plate $155–225
Rectangular "CS" two-piece belt plate $675–$1060
Large oval "AVC" waist belt plate, Alabama Volunteer Corps $585–900
Oval waist belt plate with Alabama state seal $935–1335
Louisiana waist belt plate finished in silver $550–700
Louisiana sword belt plate, stamped brass $450–650
Rectangular heavy cast-brass sword belt plate with Maryland state seal $275–400
Two-piece Maryland sword belt plate $415–620
Two-piece North Carolina state seal sword belt plate $450–715
South Carolina waist belt plate $550–700
South Carolina rectangular waist belt plate $450–550
Two-piece South Carolina plate, cast brass $450–550
Texas large oval waist belt plate with star device $375–590
Texas stamped-brass rectangular sword belt plate $300–400
Virginia waist belt plate, cast brass, state seal $400–500
Virginia waist belt plate, cast brass, "Virginia" $300–430
Virginia waist belt plate, stamped brass without solder $350–500
Two-piece Virginia sword belt plate $385–515

Artillery Equipment

Tar bucket	$20–30
Gunner's calipers	$75–100
Gunner's level	$75–100
Gunner's perpendicular	$75–100
Friction primer	$1.50–2.00

Shells

All types	$10–200

Bayonets

Colt revolving rifle bayonet, model 1855, stamped "345," plus scabbard	$10–14
Sharps rifle bayonet, model 1863, plus scabbard	$10–15
Spencer rifle bayonet, model 1860, plus scabbard	$10–15
Joslyn rifle bayonet, model 1862, plus scabbard	$25–35
Spencer navy rifle bayonet, model 1863, marked on blade "Ames MF Co."/"Chicopee"/"Mass," plus scabbard	$20–30
Confederate Hall rifle bayonet, plus scabbard	$15–20
Richmond rifle/musket bayonet, plus scabbard	$15–20

Bottles

Beer	$8–12
Bitters	$5–12
Ginger beer	$6–12
Inkwells, umbrella type	$10–40
Inkwells, cone type (clay)	$10–40
Medicine	$3–10
Whiskey	$8–15
Whiskey flasks	$25–150

Bullets

U.S., .54 caliber	$0.35–0.50
U.S., .58 caliber	$0.30–0.50
C.S., .58 caliber, six rings	$4.00–5.50
Enfield Pritchett	$0.60–0.90
Enfield pattern, plug cavity	$0.70–1.00

Wilkinson bullet, .54 caliber	$5.00–7.00
Sharps, .44 caliber, slanting breech, ringtail	$6.50–9.50
Sharps & Hankins carbine, .52 caliber	$3.50–7.00
Smith carbine, .50 caliber	$1.00–1.85
Sharps, .52 caliber, hole in base	$0.35–0.50
U.S., .69 caliber, rifled musket	$0.50–0.85
Sharps and Hankins, .52 caliber	$4.00–7.75
Maynard, .36 caliber	$5.30–9.00
Maynard, .50 caliber, flat nose	$2.50–4.35
Maynard, .50 caliber, pointed nose	$3.00–5.10
Spencer, .52 caliber	$5.90–9.40
Burnside . 54 caliber, Poultney's Patent	$1.80–3.50
Spencer, .50 caliber	$0.50–1.00
Revolver, .44 caliber	$0.85–1.35
Colt revolver, .44 caliber	$0.50–0.75
Henry rifle, .44 caliber	$2.00–3.00
Colt dragoon, .44 caliber	$0.30–0.50
Revolver, .31 caliber	$0.80–1.30
Round ball, .31 caliber	$0.50–0.85

Cartridges (Metallic Case)

Maynard carbine, .50 caliber, pointed	$4.75–8.50
Spencer, .52 caliber	$0.85–1.50
Henry, .44 caliber	$3.25–5.00
Burnside carbine, .54 caliber	$3.65–5.65
Henry, .44 caliber, snub nose, stamped "H"	$5.00–7.00

Bullet Molds

Iron mold for .36 Colt revolver	$15–20
Brass mold for .36 Colt revolver	$18–25

Buttons

All types	$10–1000

Canteens

All types	$10–45

Spurs

All types	$3–25

Knives (U.S.)

Bowie knife	$35–75
with sheath	$65–100
Bowie knife, inscribed	$50–85
with sheath	$70–125

Knives (Confederate)

Presentation knife	$100–300
with sheath	$125–400
Bowie knife	$65–100
with sheath	$75–125
Bowie knife, inscribed	$85–275
with sheath	$125–375
D-Grade Bowie knife	$100–150
with sheath	$200–350

Chapter 10

Selling Your Treasure

When you find some valuables and determine their value, you'll probably want to cash in your loot. Though minerals and gems will have a ready market almost everywhere, relics and coins will be a little more difficult to convert to cash if you expect to get their full collector's value. This chapter, however, will give you a head start in locating the markets you're interested in.

COINS

Since coin hunting is by far the most practiced field of treasure hunting, let's take a look at some of the potential markets for old coins. Of course, you can take your finds to any bank and cash them in if you prefer, or you can spend the coins if the face value is all that you're looking for. However, many coins are worth many times their face values.

In order to get top prices for your coins, offer them for sale to the highest bidder. This method offers reasonable certainty of getting the top market value.

You can start looking for potential bidders by placing an ad in your local newspaper or in magazines read by collectors. *Coin World*, P.O. Box 150, Sidney, Ohio 45365, is published for coin collectors, mostly specialists in U.S. coins or related numismatic items such as medals, tokens, and paper money. It would be a good magazine to advertise your finds in. Other magazines would include *American Collector,* 13920 Mt. McClellan Blvd., Reno, Nevada 89506; *Collectors News*, P.O. Box 156, 606 8th St., Grundy Center, Iowa 50638; *Today's Coins*, P.O. Box 919, Kermit, Texas, 70745; and the *National Hobbyist,* 805 North First St., McGehee, Arizona 71654.

Antique shops will usually handle coins—as well as any other items you want to sell—on a consignment basis; that is, you assign the item(s) over to them as your agent, and if the item is sold they'll mail you a check for the amount received less a normal fee, usually about 10% to 25%. In the case of flea markets, you usually rent space—say, $5 per showing—and then you can display your coins, relics, junk, etc., on the flea market grounds. Those who collect coins always visit flea market showings to have a look at the items displayed. You set the price on your items, but a lot of price negotiations take place at showings like these.

There are also coin shows held all over the United States. Collectors take their coin collections to these shows and also bring along coins they're willing to trade. Much trading goes on at these shows so they would be a good place to sell your coin finds for top dollar. Before attending these shows, however, make sure you know what you're talking about. Make sure you know the current value of the coins that you want to sell or trade. The main point to remember is don't be in a hurry to sell your coins. Bargaining pays off.

If you are in a hurry to sell your coins for some ready cash, try one of the coin dealers advertised in the classified sections of many magazines. There may even be a hobby shop near your home that may buy all your finds in one lump deal. Of course you'll have to settle for less money if you sell your coins this way. Try your local gift shops also. Owners of shops like these are always in the market for profit-making items.

OLD BOTTLES

There are many places where you can sell your old bottles. Most of the antique shops throughout the country now carry a line of antique bottles and will either buy your bottles directly (at about 50% of their collector's value) or else will take them on consignment. Flea markets also have several displays of antique bottles for sale or trade.

Don't overlook your local newspaper as a means of contacting local collectors. The advertising rate will more than likely be reasonably low and you may even learn that you have a collector living within a block or so of your home who will buy your entire bottle collection at top dollar.

National magazines like *The Old Bottle Magazine*, Box 243, Bend, Oregon, 97701, are also good places to run ads. Glance through some other likely magazines at your local newsstand; maybe they also run ads concerning old bottles.

When you visit antique shops, ask the owner if he has any antique trade journals on hand. Jot down the names and addresses of each journal that is concerned with old bottles. Once you are a subscriber to one of these publications, you can keep an eye on the booming bottle market prices. You will learn just what is in demand so you can be on the lookout for particular bottles on your next treasure-hunting outing.

Of course there are bottle shows or sales, just as there are for almost all other collectibles.

GEMS AND MINERALS

Everyone knows that he can easily sell gold or silver just about anywhere in the world, but before you attempt it, check on all regulations pertaining to the possession and sale of such minerals. If you're a newcomer at this, it's recommended that you subscribe to such magazines as *Gems and Minerals*, P.O. Box 687, Mentone, California, 92359; *Lapidary Journal,* P.O. Box 80937, San Diego, California, 92138; *Rock & Gem*, 16001 Ventura Blvd., Encino, California, 91436; and *Rockhound*, P.O. Box 328, Conroe, Texas, 77301. Besides finding market information in these magazines, you'll also get a lot of information about gems and minerals—like instructions on how to collect them and step-by-step how-to articles on lapidary and jewelry making.

There are many other sources of information for the beginning rockhound—libraries, books, reports of state and federal agencies, mineral dealers, mineral societies, museums, schools, and collectors. All are a possible source of market information.

The following dealers and suppliers may also help you sell your finds:

John S. Albanese, P.O. Box 221, Union, New York
Allen's Minerals, McCoy, Colorado
Althor Products, 2301 Benson Ave., Brooklyn, New York
American Optical Company, Buffalo, New York
American Phenolic Corp., Chicago, Illinois
Bausch and Lomb, Inc., Rochester, New York
J.E. Byron, 634 Highland Ave., Boulder, Colorado
Jim Carnahan, 9531 Mina Ave., Whittier, California
Covington Lapidary Eng. Corp., Redlands, California
Edmund Scientific Co., Barrington, New Jersey
Graf-Apsco Co., 5868 Broadway, Chicago, Illinois
Herrick Micro Projector, 2436 Holmes St. Kansas City, Missouri
Highland Park Manufacturing Co., South Pasadena, California
Robert Lobel, 28265 Beatron Way, Hayward, California
The Micromart, P.O. Box 3007, Compton, California
Micromounts Unlimited, 2525 Rivermont, Lynchburg, Virginia
Minerals-West, 656 S. Hendricks Ave., Los Angeles, California
Minerals Unlimited, 1721 5th St., Berkeley, California
Harry Ross, 61 Reade St., New York, New York
Harry Sering, Indianapolis, Indiana
Shortman's Minerals, 10 McKinley Ave., Easthampton, Massachusetts
Southwest Scientific Co., P.O. Box 457, Scottsdale, Arizona
Speedy Products, Inc., Richmond Hill, New York
Sturm and Smith Publishers, Tucson, Arizona
Swift Instruments, Inc., P.O. Box 562, San Jose, California
Terry's Lapidary, 3616 E. Gage Ave., Bell, California
Unitron, 66 Needham St., Newton Highlands, Massachusetts
Ward's Natural Science Estab., Rochester, New York
William H. Yost, 9802 Redd Rambler Dr., Philadelphia, Pennsylvania

TOOLS AND RELICS

When searching around old buildings for treasure, you may find such items as axe heads, kerosene lamps, and other tools and relics. The best place to sell items like these is a flea market; the second best place would be an antique shop. If you can't find either around your home, subscribe to *Flea Market Quarterly*, Box 243, Bend, Oregon, 97701. Ads in this magazine should certainly turn up some potential markets for your finds.

CIVIL WAR RELICS

The interest in Civil War relics has been on the rise for nearly a century, but only recently has the collecting of these relics really boomed. It seems like everyone and his brother, especially those living around battlegrounds, are collecting relics from Civil War days—Minie balls, sabers, bayonets, spurs, uniforms, papers, even ink bottles.

It is a hard matter indeed to drive very far from any of the better known Civil War battlefields and not see several dealers in relics. These same dealers are also a potential marketplace for your finds. But you'll probably get a better deal if you try some of the shows or advertise in local newspapers.

In the magazine *North South Trader*, 8020 New Hampshire Ave., Langley Park, Maryland, 20783, you will find page after page of ads concerning the buying and selling of Civil War relics. Besides items like muskets, swords, buckles, and buttons, you will also find want ads for Union and C.S.A stamps, envelopes, and letters.

Every year there are a lot of shows where only Civil War relics are allowed to be exhibited. These are good places to rub shoulders with collectors and obtain the highest prices possible for your relic finds. The Shenandoah Valley Relic Hunters Association recently held their semi-annual Civil War relic show with 50 tables of quality Civil War memorabilia for sale and show. Tables at such shows are for rent; that is, you pay to rent a table on which to display your wares. And there are always plenty of collectors around to sell to.

Try running a few ads in collectors magazines. In many of these magazines the classified advertising rates are only about $0.25 per word. All ads should be mailed at least one month before publication date.

Chapter 11

Caring For Your Finds

This chapter offers suggestions on preserving and caring for coins, relics, and other valuables that you may find. Using the following techniques will help make your finds last longer and sell at higher prices.

COINS

When you return from a day of "coin shooting", you should immediately remove all soil, sand, and corrosion from the coins you have found. Do not, however, use any *abrasive* chemical on them. Collectors prefer coins whose surfaces are clean but unmarred.

To clean a coin, use an old toothbrush with ordinary dishwashing detergent and warm water. Add the detergent to the water. Dip the coin in the water, take it out, then scrub it with the toothbrush. Usually about one teaspoon of detergent to a quart of warm water will suffice. But for stubborn grime and corrosion soak each coin individually in undiluted detergent then brush each coin carefully with the toothbrush.

After cleaning, rinse all coins in clear water, then isopropyl (free of impurities) alcohol. When dry they are ready for checking, grading, and mounting. The checking and grading can be done best with the use of an inexpensive magnifying glass.

Coins can be stored in plastic tubes with screw-on caps, or they can be inserted in small square cardboard holders (one for each coin). Some collectors use cardboard coin albums. You can buy them from any coin dealer. Some collectors may even display their collections in picture frames.

You may be able to get more for some coins by mounting them into pieces of jewelry like key chains or clasps. Belt buckles made from silver dollars would bring 10 times the coins' face value. You are probably beginning to get some ideas of your own by this time, but remember, if you plan to sell the more valuable coins to collectors, it is best to just clean the coins carefully and store them as described previously.

OLD BOTTLES

People from all walks of life are turning to the fascinating hobby of bottle collecting. Though most of these bottles are only worth a dollar or two, some have brought over $200 each. In general, all bottles should be carefully cleaned, but extra care should be taken for those with paper labels because the bottles will be worth more with the labels intact. Do nothing that will in any way damage the label.

If the bottle has no paper label, it can be washed in warm water and regular dishwashing detergent, using a bottle brush if necessary. If a paper label is still attached, wash the inside of the bottle with cold or lukewarm water (with detergent), using a funnel to assure that no water spills down onto the label. Then take a damp cloth and wipe off the dust on the outside, but use no liquid which might damage the label or cause it to come off.

Some bottles are found with all or part of their original contents still in them. In some instances, such finds make the bottle worth more, so don't remove the contents until you have found out if it's worth anything. Just wipe off the outside of the bottle with a damp cloth to remove the dust; leave the interior alone. If the collector who buys the bottle wants the contents removed, he can do it himself. One recent find of this nature is shown in Fig. 11-1. It's a glass-top jar containing some pickled minnows (used for fishing bait) and it's estimated to be about 50 years old.

Serious bottle collectors often work for hours to restore damaged bottles. They clean oxidation away with acid, repair broken parts with fiberglass or epoxy cement and sometimes even try to restore labels.

Fig. 11-1. This glass-top jar of pickled minnows is about 50 years old.

The first step in restoring bottle labels is to identify the bottle and locate an authentic bottle label to match it. Then you can use a high quality copying machine (Xerox, IBM, etc.) and copy the original label. If you want an aged appearance, merely soak the copy in coffee or tea for a short period of time.

In some cases, a badly faded label or one that is recessed somewhat cannot be reproduced effectively. If this is the case, try a camera with a closeup lens using high contrast film. Make print from the negative to match the size of the original label. Copy the black and white print with a copying machine. Use the correct color paper. After this process, you can stain the label with tea or coffee.

GEMS AND MINERALS

Most gems are made into jewelry. But often the hobbyist displays his gems and minerals in micromounts.

A wide variety of materials have been used to make bases for micromounts, but corks, balsa wood, toothpicks, bristles, pins, and tacks seem to be the most popular. The process involves selecting the specimen, preparing the pedestal and container, trimming the specimen to size, and cleaning the specimen. Then it is ready for mounting on a pedestal.

In general, large specimens should be mounted on pedestals having a base about the size of the specimen. Smaller specimens are usually mounted on toothpicks and the like. After mounting, specimens are placed in a holding box and labeled and cataloged according to each specimen's classification.

The book *The Complete Guide to Micromounts* (Gem Books) by Milton L. Speckels is a good guide for those interested in this hobby.

CIVIL WAR RELICS

Give each relic you find an indentification number and make notes of where you found it. If a map of the area is available, mark the approximate spot where the object was found. The fact is, some people are willing to pay more money for a relic if they know where it came from, especially if it came from an important battlefield.

For example, you can go in any of the souvenir shops around old battlefields and see Minie balls and other types of bullets on sale, but they are probably going to be in a tin bucket—all mixed together—with no one knowing where any of them came from. On the other hand, suppose you walked into a shop with two or three Minie balls attached to an attractive card that states, "Fifty-eight caliber Minie balls found on the Milford, Virginia, battlefield (conflict on June 8, 1862) in front of the trenches occupied by Harris' 2nd Brigade. Since these bullets show no sign of firing, they were probably dropped by the troops occupying these trenches." Wouldn't you rather own the bullets just described than ones you know nothing about?

Chapter 12

Building Your Own Treasure-Hunting Equipment

The circuits described in this chapter are for treasure hunters who want to save from 20% to 60% on the cost of some of their equipment. A few of the items discussed here—like the transmitter-receiver metal detector and the electronic depth sounder—can be purchased in kit form from electronic supply houses. Other items are metal detectors that can be built from scratch with components lying around your home.

All of these circuits have been chosen for their reliability, ease of assembly, and low cost of the electronic components. When properly constructed, these projects will yield high-quality performance.

If you're new at constructing electronics projects, perhaps you should read up on basic electronics. TAB Books offers several volumes on the subject, and the instructions included with most of the kits include a basic course in electronics.

The tools needed to build any of the kits in this chapter include a pair of needlenose pliers, a pair of diagonal wire cutters, a small pencil-type soldering iron, two regular and two Phillips head screwdrivers, a pair of all-purpose pliers, and a small electric drill. Other tools that will come in handy are metal-cutting dies, a small bench vise, and a standard AM broadcast radio for monitoring detector frequencies. You will also need a selection of small bolts, nuts, and screws for the metalwork involved.

If this is your first home electronics project, heed the following basic rules. Do not begin the project until *every* component needed is on hand. Do not set a deadline for having the project completed—take your time and enjoy the project. Before soldering any component, double-check your connections to make certain that they are correct. When you get tired, take a break.

Many hobbyists prefer to complete the mechanical work first, before tackling the circuitry. The electronic wiring may seem the more difficult of the two, but the presence of a completed mechanical shell to house the components could be an added incentive to keep you going.

The majority of home electronics projects that do not operate correctly upon completion fail to do so because of improper soldering techniques. Therefore, make certain that all your soldering joints are good. In general, this is accomplished by applying the soldering tip to the joint and not to the solder. When the wire joint becomes hot enough, the solder will flow onto it, leaving a shiny film when cool. A dull or jagged solder joint is an indication of an improperly soldered joint that is certain to present troubles in the future.

When building the metal detector kits, the pickup coil should be wound on a structurally stable form, such as a Quaker Oats cardboard container. The coil itself may then be mounted on a plastic Frisbee or on several stiff paper plates cemented together and then varnished.

Mechanical stability is very important when constructing any of the projects because poor stability will result in drift of the oscillator frequency and poor results overall. Therefore, after your project is completed, give the unit a firm shake, observing all circuit components. If anything moves, dab several drops of epoxy cement on the "guilty" component to secure it more firmly. The finest metal detectors on the market today are those that boast of stability as well as advanced circuitry.

SIMPLE METAL DETECTOR

The schematic wiring diagram in Fig. 12-1 shows a search coil (L_1) consisting of 22 turns of No. 22 enameled magnet scramble wound on a six-inch form. The center tap should be made on the 11th turn. The search coil of this Hartley oscillator circuit is then mounted on a search head consisting of a plastic Frisbee or paper plates

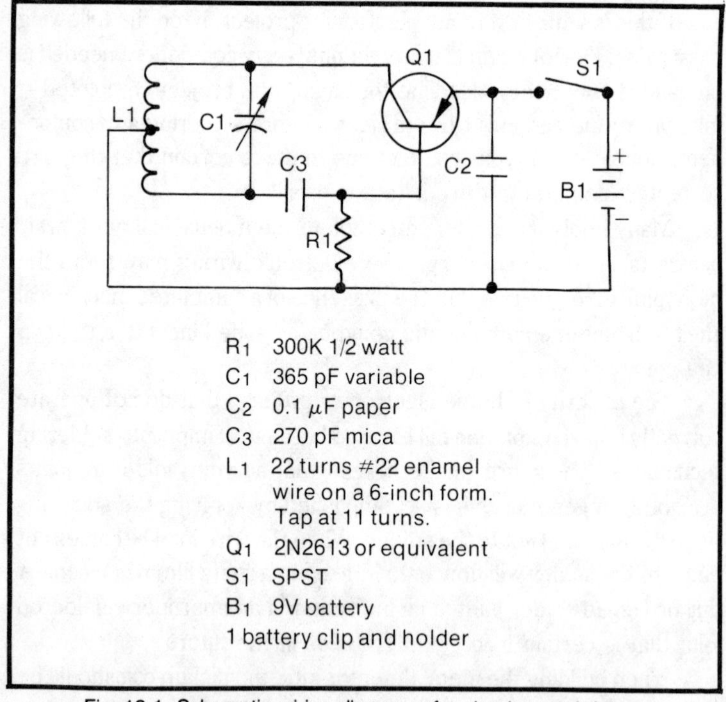

Fig. 12-1. Schematic wiring diagram of a simple metal detector.

(stacked and varnished) connected to the circuit components by means of a three-conductor insulated cable.

After the circuit is completed and the components are mounted as shown in Fig. 12-2, the detector is used in conjunction with a small transistor radio. Tune the radio between two local stations, then

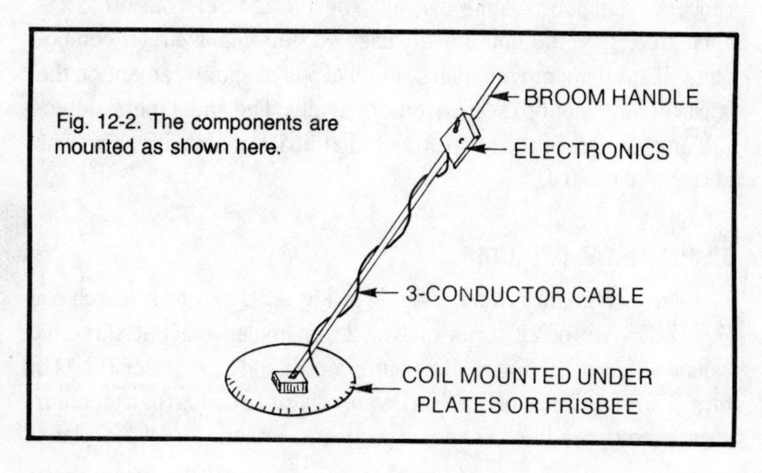

Fig. 12-2. The components are mounted as shown here.

adjust the variable capacitor (C_1) until a "beat" whistle is heard. When the search head is passed over a metallic object, the pitch of the whistle will change either up or down. The amount of change will depend upon the size of the object located; that is, the larger the object, the more significant the change in the beat tone or whistle.

Tune this detector (or any other detector) outside the house, away from all metallic objects. Wet grass may also affect the signal, giving false tones.

COMPACT METAL DETECTOR USING HARTLEY OSCILLATOR

This is another simple metal detector circuit (Fig. 12-3) that uses a small portable transistor radio to detect and amplify the beat generated by the oscillator. This detector uses an NPN transistor instead of a PNP transistor. In addition, nearly all the component values have been changed to offer the builder the possibility of using up some old electronic components lying around the house.

Probably the most significant change is the smaller search coil which consists of 28 turns of No. 22 AWG enameled magnet wire wound on a four-inch form and tapped at 15 turns. The small coil of

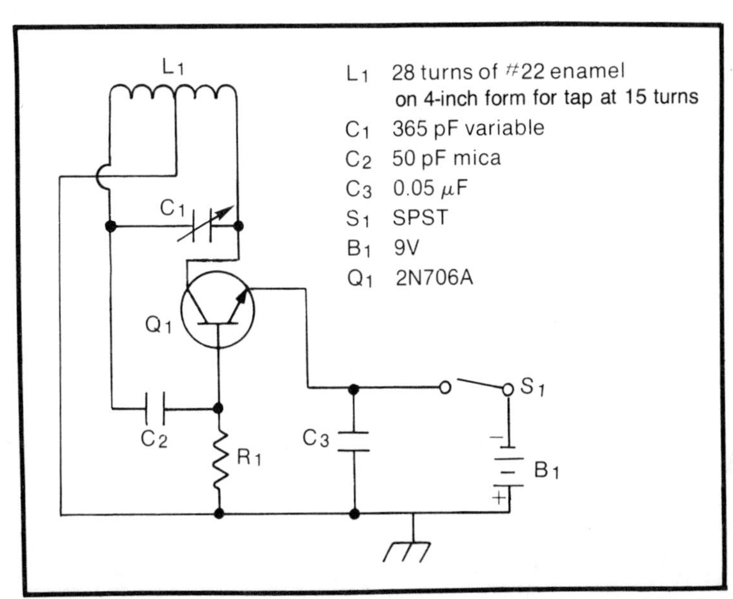

Fig. 12-3. Schematic wiring diagram of a compact metal detector similar to the one in Fig. 12-1.

this Hartley oscillator model should provide greater sensitivity on small items such as coins and rings, but some depth will be lost.

Mechanical details of this detector are exactly like the ones described for the circuit in Fig. 12-1. Scramble wind the coil on a form (a Quaker Oats container is perfect) and construct the circuit components. Connect the coil leads to their proper location on the circuit board and secure all items on a wooden pole or aluminum tubing as shown in Fig. 12-2.

COMPACT METAL DETECTOR USING COLPITTS OSCILLATOR

A small portable transistor radio is necessary to operate the circuit in Fig. 12-4. In this circuit, however, the search coil (L_1) has

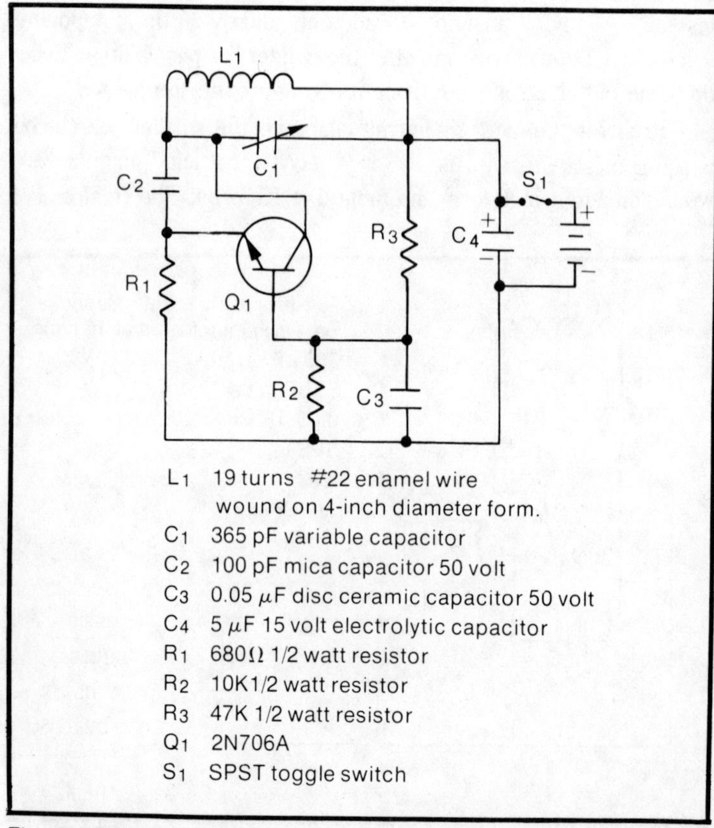

Fig. 12-4. Schematic wiring diagram of a compact metal detector using a Colpitts oscillator.

no center tap. Instead, a Colpitts oscillator circuit is used which does not require a tapped coil. With a small four-inch coil and slightly superior design this metal detector circuit should be somewhat more sensitive than the two previously described. However, when testing this particular model, it was found that the added sensitivity also picked up false signals from wet grass and sand more readily. With practice the user should be able to tune out these signals or at least be able to distinguish between false signals and true finds.

Exact values of components (except the battery) are not critical and each may be varied as much as 20% without any adverse effect. When completed, tune the transistor radio between two local stations or to one side of one station and adjust the variable capacitor until a low beat whistle is heard. Keep the search head as close to the ground as possible (without touching it) and search for buried metallic objects by swinging the search head back and forth over the ground in front of you as you walk. A change in the beat whistle tone indicates a find.

ADD-ON AUDIO-VISUAL INDICATOR

The circuit in Fig. 12-5 is designed for use with complete self-contained metal detectors with headphones to hear the beat. This circuit, when completed, is connected to the headphone jacks (B and A). You then have both an audible speaker to hear the beat as well as a visual aid, a 0–50 microampere meter.

Construction of this circuit is not difficult, but make certain that you wire the circuit to the correct polarity as shown.

HEATHKIT TRANSMITTER-RECEIVER METAL DETECTOR

The Heathkit Model GD-348 Deluxe Metal Locator is a highly sensitive instrument for finding buried or hidden metals. Because it is simple to operate, anyone—amateur, hobbyist or professional—can use it to pinpoint metal objects with accuracy. This lightweight unit is designed for comfortable handling to provide many hours of exciting treasure hunting, even by youngsters, and can be built in an evening or two by almost anyone.

Separate *sensitivity* and *null* controls are provided for unlimited versatility over varying geological locations. A speaker and a meter are provided to give you both a visual and an audible indication when

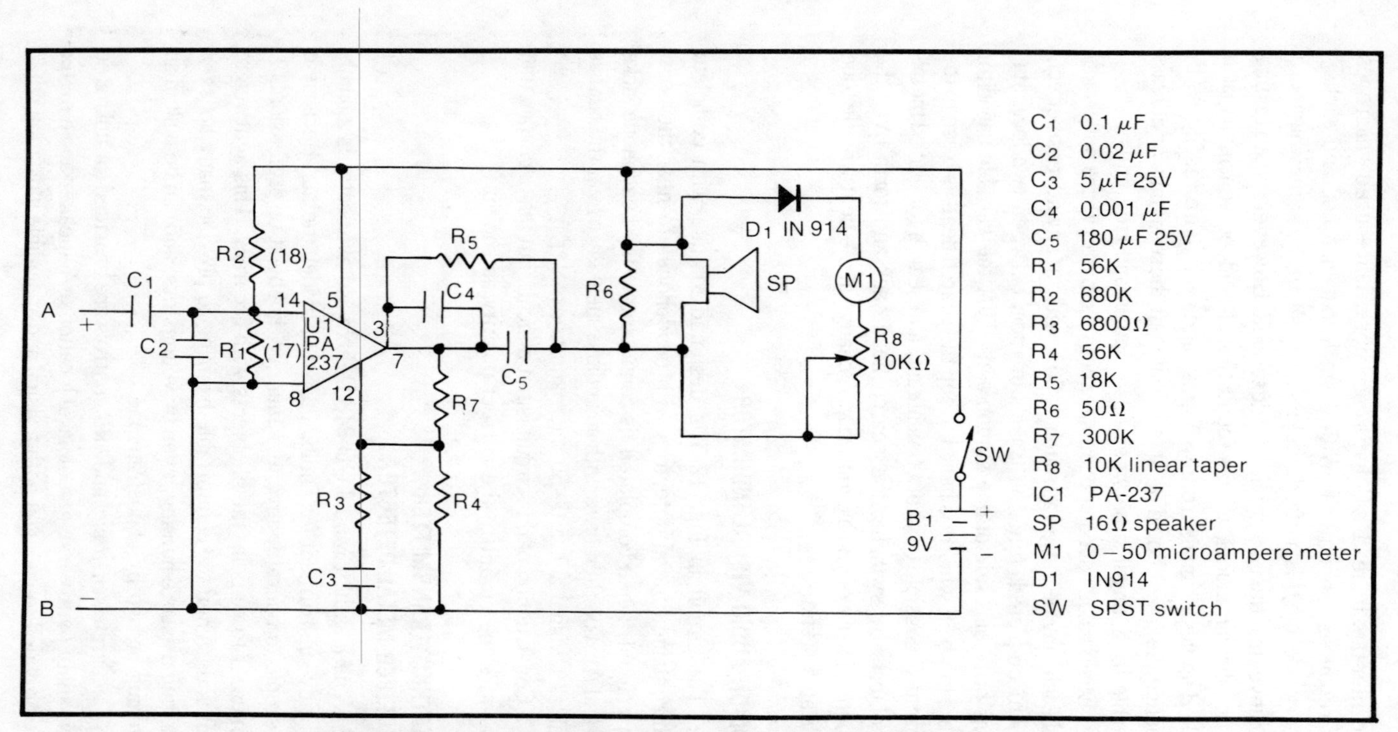

Fig. 12-5. Schematic wiring diagram of an add-on audio-visual indicator.

metal is detected. A jack for headphones is also provided for use in areas where the surrounding noise level is high. In addition, the search coil housing is designed for submersible operation in up to two feet of water (more if you detach the search head and install a longer cable).

Rugged construction and versatility of operation make this deluxe metal detector dependable and a pleasure to use. New uses will be found each day to add to your enjoyment and adventure.

Once the kit is partially assembled (according to the instructions provided with the kit), you will want to make the following initial adjustments (refer to Fig. 12-6):

1. Position the Deluxe Metal Locator so the search coil is positioned as shown in the inset drawing. Insure that the search coil is away from any metallic objects.
2. Rotate the *null* control to the center of its rotation.

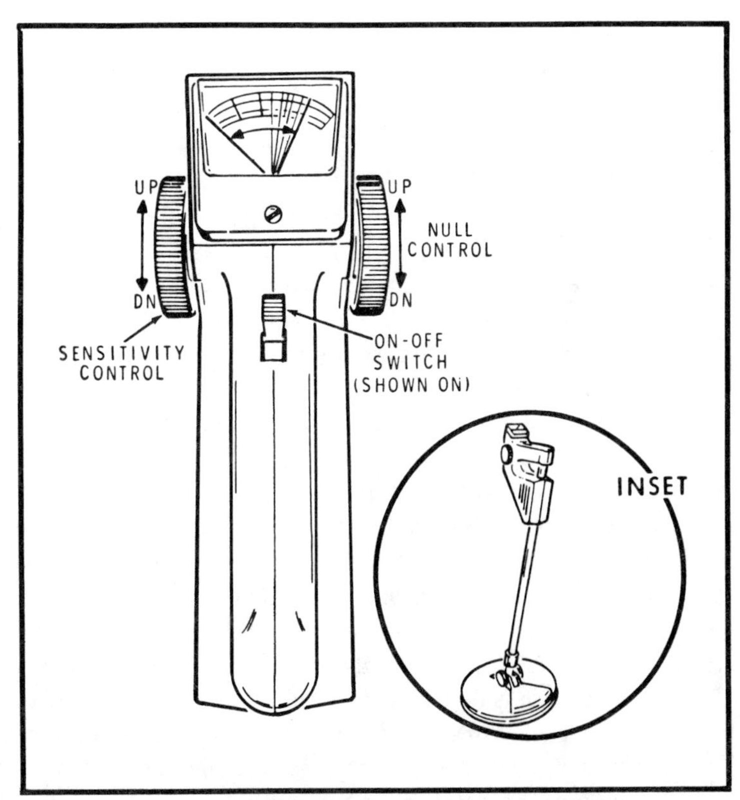

Fig. 12-6. The controls on the Heathkit Metal Locator.

3. Slide the *on-off* switch to the *on* position. You may or may not hear a tone from the speaker. *Note:* The left handle half is lettered *max*, indicating the control direction for maximum sensitivity. *Up* and *down* will be used as a convenient reference in some of the following steps instead of the *max* indication.
4. If you do *not* hear a tone, rotate the *sensitivity* control *down* until a meter indication of six is obtained. If you hear a tone, rotate the *sensitivity* control either *up* or *down* until you obtain a meter indication of six.
5. Slowly rotate the *null* control in either direction until you obtain a minimum meter indication (null). *Note:* It is normal if the meter needle drops and remains at zero during some of the control rotation.
6. Rotate the *sensitivity* control *down* for a meter indication of six and then rotate the *null* control for a minimum meter indication. When rotation of the *null* control just *increases* the meter indication, the best null has been reached.
7. Rotate the *sensitivity* control so the pointer indicates halfway between zero and two on the meter.
8. Slide the *on-off* switch to *off.*

After the detector is fully assembled, give it the following operational check (refer to Fig. 12-7):

1. Place a dime on a nonmetallic surface where the search coil can be passed over it. The most sensitive spot on the search coil is just forward of the shaft thumbnuts. When this spot on the search coil is directly over buried metal the strongest indication will be obtained, thus pinpointing the location of the metal.
2. Slide the *on-off* switch to *on*.
3. Read just the *sensitivity* and *null* controls while holding the search coil about two inches above the nonmetallic surface.
4. Hold the search coil approximately two inches above the nonmetallic surface and gradually pass it over the dime. Figure 12-7 illustrates the proper meter indication as a small coin is detected. The audio tone will also increase. Headphones may be used to provide a more sensitive means of detecting this increase. The speaker is automatically disconnected when headphones are used.
5. Slide the *on-off* switch to *off.*

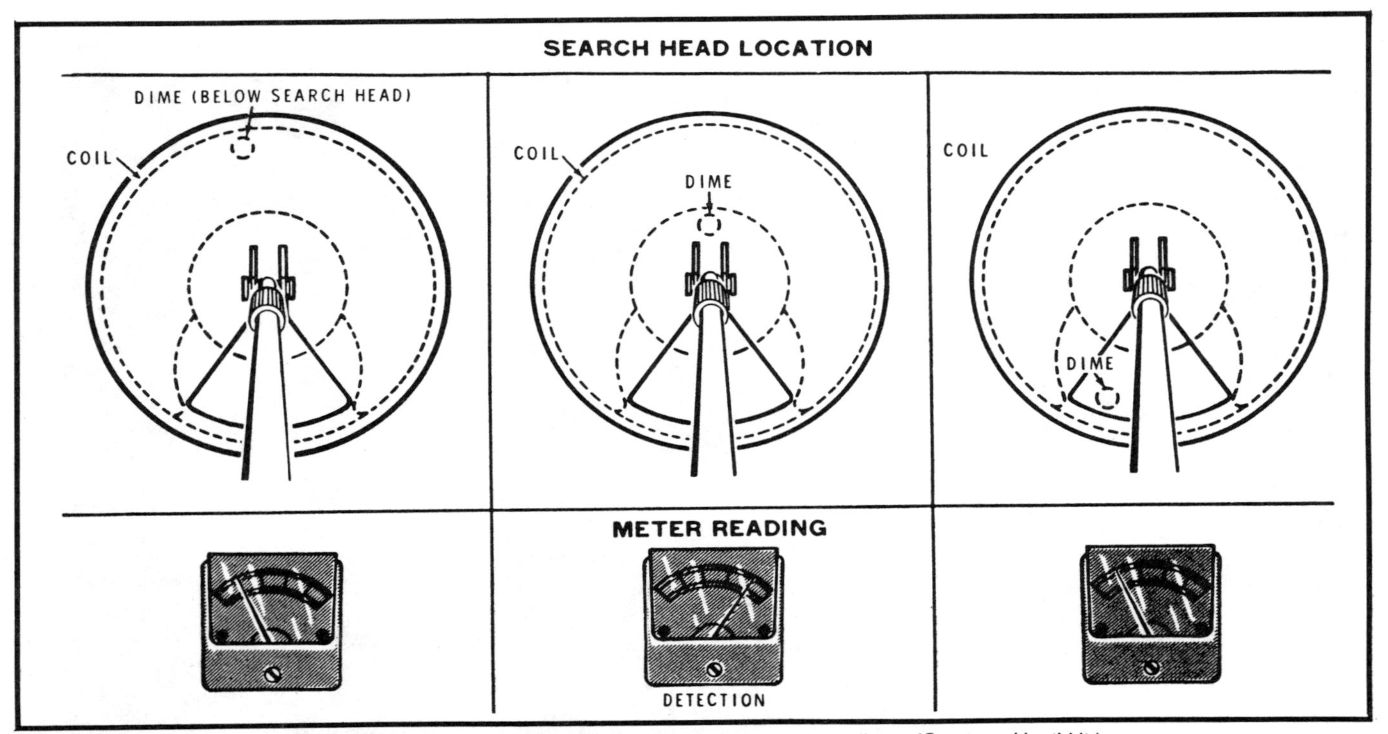

Fig. 12-7. Search head location and corresponding meter readings. (Courtesy Heathkit.)

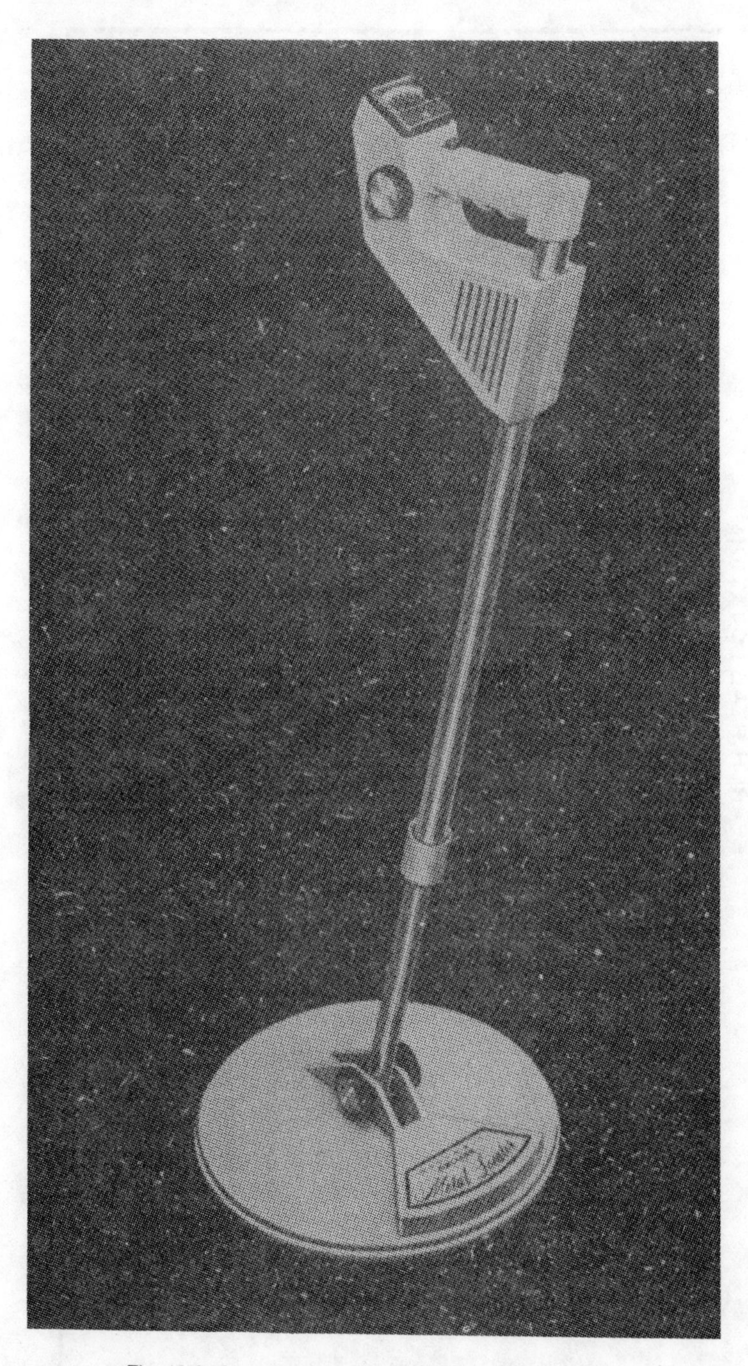

Fig. 12-8. A completed Heathkit Deluxe Metal Locator.

After this operational check, again make the adjustments as previously described.

This locator is designed for easy operation, but it may require a small amount of practice before you can get the best results and locate buried metal accurately. Read the operation section of the accompanying manual carefully to obtain a greater understanding of the locator.

Locating small metal in tall grass, in wet ground, and especially in wet gravel or sand may prove difficult. For best results the ground should be fairly dry.

The search coil may be pivoted 90° on the shaft and folded flat against it (Fig. 12-8). Forcing the search coil to rotate when the shaft is not fully collapsed will damage the shaft liner. When the shaft is fully collapsed the metal detector can be carried in a carrying case that comes with the kit.

When you search for small metal objects, such as coins, or for deeply buried metal, it is best to use the signal meter which provides a more sensitive indication than the audio signal.

Fig. 12-9. A completed Heathkit depth sounder.

Practice finding various sizes and types of metal you have planted. The better you become acquainted with your Deluxe Metal Locator, the easier it will be for you to find buried metal.

Remove the battery when the Deluxe Metal Locator is to be stored for a long period of time.

HEATHKIT DEPTH SOUNDER

Another treasure-hunting aid available in kit form from Heath Company is a chart-recording depth sounder (Fig. 12-9). Again, step-by-step instructions are given to enable the hobbyist to construct this device at home.

The bottom recording accuracy makes the chart recorder extremely valuable in "seeing" ship wrecks as well as spotting schools of fish. It will also prove valuable as a navigation aid if you need to follow the coast lines or operate a boat in harbors, in fog or at night.

Chapter 13
Getting Better Results With Treasure-Hunting Equipment

Upon reading the title of this chapter your first thoughts are probably going to be about rewiring metal detector circuits in order to make them more powerful. Though it is certainly possible to soup up some detectors, most have been carefully designed by electronic engineers and are difficult to improve upon—at least by the average treasure hunter. Still, there are many other ways to improve your chances of finds without getting too technical.

ELECTRICALLY CHARGING AN AREA

One treasure hunter uses a surplus 100V hand-cranked generator to electrically charge all metallic objects in a given area so they will be easier to locate with a metal detector. Many of these generators are available for less than $10. Merely attach two 12-foot leads (#12 or #10 AWG wire) to the terminals on the generator, saw one eight-foot copper-clad ground rod in half (making two four-foot lengths), and attach one of the leads to one of the rods and the other lead to the other rod. Before cranking the generator handle, the rods are driven into the ground at intervals depending upon the soil conditions. The result will be the charging of all metallic objects between the two rods.

In order to test this procedure, bury various metallic objects at different depths until your detector will barely pick up the objects. Then bury them a bit deeper so that your detector will not pick them up at all. Insert the rods (electrodes) into the ground between the objects and crank the handle of the generator. Now see if your detector will pick them up.

SEARCH COILS

Proper coil selection is very important to the success of any treasure hunt, and this is a way that the treasure hunter can increase the sensitivity of his equipment. A small coil is generally more sensitive than the larger ones. The small ones have a dense electromagnetic field which will react strongly to small metal objects, but these objects, as a rule, must be relatively close (near the surface) in order to be detected. The larger coils produce larger and deeper detection fields but require somewhat larger targets in order to permit deep detection.

You can see then that small objects or objects near the surface require only a small search coil. On the other hand, hidden caches will probably be buried much deeper and better results can usually be obtained with the larger search coils.

A complete line of search coils will usually be available for your metal detector. The following list is a summary of the search coils available from Garrett Electronics, 2814 National Drive, Garland, Texas, 75041. If you need any further help in selecting a metal detector or a search head, feel free to contact them for assistance. The crew at Garrett have had years of training and experience in all phases of treasure hunting and they can offer you much valuable advice on the subject.

Three and one half-Inch Search Coil. This size search coil is excellent for coin hunting as well as detecting small nuggets. It is used with the BFO detectors. It locates and traces gold placer deposits and small ore stringers. It is good for scanning walls, ceilings, floors, and small, confined areas. Garrett Electronics recommends this size coil for nugget hunting and metal/mineral identification. It's just the right size for beginning BFO operators who have not mastered the larger coils.

Five-Inch Search Coil. This size is good for coin hunting since excellent response is obtained when detecting even the smallest of coins. It can be used successfully for detecting larger nuggets, ore samplings, and shallow veins. Wall, floor, and ceiling scanning can be easily accomplished with this size. It's also good for shallow relic searching when a larger coil is not available. This size is normally used with the BFO detector.

Eight-Inch Search Coil. The eight-inch coil is a good general purpose coil for either the BFO or TR detectors. It is small enough to give reasonably good response on buried coins and large enough to do a good job of detecting small relics buried within the first foot or two of the surface. If a small coil is available, the beginning treasure hunter should use it when first learning to search for coins. Then, after becoming familiar with his detector, he can advance to the eight-inch coil for coin hunting. The reason for this is that the eight-inch coil (when used with BFO detectors) will not give as definitive a signal response to coins as the smaller coils, but the larger coils have a wider sweep path which allows faster coverage of an area. The eight-inch shielded 4B type search coil used on the TR detectors can act as a *standard* TR search coil as well as a *discriminating* TR search coil. Figure 13-1 shows a Garrett TR detector with the eight-inch 4B search head.

12-Inch Search Coil. This size is designed for cache hunting with the BFO detector. A cache can be detected down to four feet

Fig. 13-1. A Garrett TR detector with an eight-inch 4B search head.

Fig. 13-2. A treasure hunter operating a Garrett Master Hunter TR metal detector with a 13 × 16-inch coil.

with this size. This size is not recommended for coin hunting, but it can be used successfully for the rapid scanning of old houses and ghost towns.

13 × 24-Inch Search Coil. This size coil was designed for use with BFO detectors and is an improved deep-seeking coil that detects relics and ore veins. It can search deeper than the 12-inch coil. It will also double the scanning speed of the 12-inch coil because of its elongated wide-profile search pattern.

24 × 24-Inch Search Coil. For maximum depth penetration on larger objects, this coil is hard to beat, although its size makes it hard to maneuver. It is used with the BFO detectors to search for pots of treasure, chest-sized objects, and ore veins.

13 × 16-Inch Search Coil. This coil has been developed for use with the TR detector when searching for relics, buried pipes, bottle dumps, and caches in general. It can search deeper than the standard TR coils, but it does not discriminate. Figure 13-2 shows a treasure hunter operating a Garrett Master Hunter TR detector with the 13×16-inch coil. A tin can filled with coins was located at a depth of over two feet with this detector.

Double search coils (3 1/2 inches and 6 1/2 inches) are also available for use with BFO detectors. Two completely separate coils are combined in one search head to offer a versatile coin hunting pattern. The 3 1/2-inch size will give the most precise pinpointing, detecting the tiniest of objects. The 6 1/2-inch size was designed to detect the smallest coins with the maximum possible response, sensitivity, and scanning width. When you flip the switch to change from one coil to the other, you can rapidly determine the approximate target size and depth. Figure 13-3 illustrates the differences between the search patterns of the two coils. Obviously, double

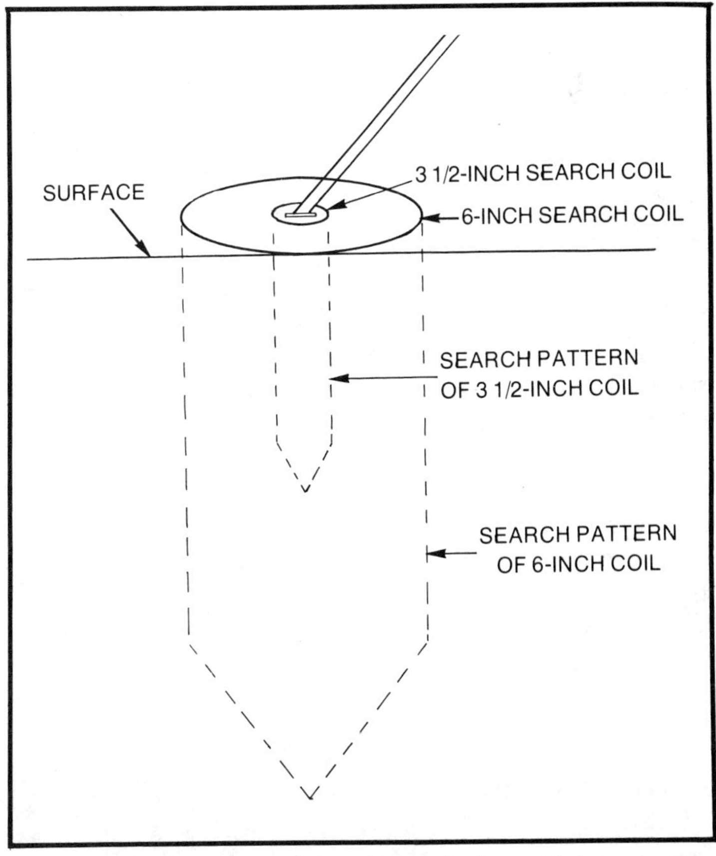

Fig. 13-3. The detection pattern of two independently operated search coils.

search coils give you all the advantages of both a 3 1/2-inch search coil and a 6 1/2-inch size. In addition, you also get the "instant change" feature. If other combinations would better suit your treasure-hunting activities, there are also 3 1/2-inch/8-inch models and 5-inch/12-inch models.

Those treasure hunters who spend much of their time searching in water will be glad to know that there are also several sizes of underwater coils available. One is a 5-inch search coil that can be lowered down into 50 feet of water to detect any size objects on the bottom or covered by a few inches of sand or mud. A coil of this size is better for well and cistern searching than a larger type. It will locate boats, motors, fishing tackle, guns, and smaller objects. A diver can maneuver the coil over the bottom while the operator in a boat monitors the control housing. A tug on the cable will alert the diver to a find.

A larger 12-inch underwater search coil (also provided with 50 feet of cable) is available and operates just like the 5-inch coil. This larger coil is designed primarily for larger object detection at a greater distance below the search coil (under sand, mud, etc.). It is not recommended for coin-sized object detection.

EARPHONES

Most commercially manufactured metal detectors have built-in provisions for plugging in a set of earphones. Many treasure hunters use a set of earphones when searching in a crowded area so as not to alert other hunters of their finds. A high-quality set of dynamic 8Ω stereo earphones with cushioned earpieces will also do wonders in boosting faint signals.

If you choose to use a set of earphones while operating your metal detector, there are two precautions that should be taken: (1) While searching in snake country, leave one earphone off one ear so that you can hear, say, a rattle from a rattlesnake; (2) Because of the power capability of the audio circuit in most detectors, use caution when first turning on the earphones because full detector power through the earphones could cause ear discomfort or even ear damage.

DISCRIMINATING CIRCUITS

Some metal detectors have twin circuits: a TR circuit for providing positive responses to all types of metals, and a discriminating circuit for distinguishing between real finds and junk (nails, pull tabs, etc). The discriminating circuit automatically "rejects" junk. Most of these detectors have controls for adjusting the discriminating circuit to any degree of "rejection," from tin foil to aluminum pull tabs. At the lesser degree of rejection, there is only a very slight degree of rejection from tin foil to aluminum pull tabs. At the lesser degree of rejection, there is only a very slight degree of rejection present (items like wire and nails are rejected). As the control is turned, however, the detector will begin to reject foil and bottle caps. The farther the control is turned in the rejection cycle, the more items will be rejected.

A meter used in conjunction with the discriminating circuit will tell you when you have passed the search head over any metal and further indicate if the find is junk or a valuable object. For example, with the detector in the discriminating mode, the needle of the meter should point approximately halfway on the scale, as shown in Fig. 13-4. When you pass over a metallic object, the needle should change to indicate whether the item is junk or a valuable find. If the

Fig. 13-4. With the detector in the discriminating mode, the needle of the meter should point approximately halfway on the scale.

needle moves all the way to the right, then the meter has indicated a find (something not rejected by the discriminating circuit). However, if the needle swings to the left, then the item is probably junk.

When adjusting your detector for discrimination, sweep the search head over a sample reject object (bottle cap, etc.) as you would when actually using the detector in the field. Don't allow the reject objects (during a test) to touch the bottom of the search coil. Hold the search coil about one inch away from the objects.

The addition of a meter to your detector will not only increase its sensitivity, but will give you full "silent" searching capabilities while allowing the use of your ears as a safety precaution in areas like snake country, along highways, etc.

COLD WEATHER DETECTING

You may spend several weekends on an old battlefield during warm weather and completely comb the entire area, finding many valuable objects. But after a while you're bound to run out of luck. But chances are if you go over this same area during *cold* weather you could possibly turn up as many finds as you did previously. Why? You already know that most substances expand when they are heated and contract when made colder. This is true with metallic objects in the ground. When they are subjected to colder weather, their molecules contract, making the substance more dense. Thus, your metal detector is able to detect them more readily.

This, of course, applies only to certain depths—depending upon the area. Most items affected by cold weather will be less than 18 inches from the surface, so the deeply buried caches will remain at ground temperature (approximately 52°F) the year round. Items like Minie balls lying between four to eight inches under the surface will be greatly affected by the cold.

WATERING DOWN

Experienced treasure hunters will tell you that when operating their metal detectors in extremely dry soil a loss in depth penetration will be noticed. Therefore, it stands to reason that if a dry area is first watered down you will be able to detect objects at a somewhat greater depth than you would otherwise. In most treasure-hunting areas, the means of watering down an area will be limited. But why not try a small area (say 10 square feet). You may have to carry water in a canvas bucket from a nearby stream, but it could pay off. The soil doesn't have to be extremely wet, merely moist.

Chapter 14

Troubleshooting Metal Detectors

Troubleshooting metal detectors is like troubleshooting any other piece of sophisticated equipment. You look at the obvious first and investigate possible causes in a logical order.

The following is an abbreviated list of symptoms with accompanying possible causes and cures:

No Speaker Volume or Sound

This problem is generally caused by either bad batteries or good batteries that are not seated correctly in their sockets. Since most of the better metal detectors have a built-in battery checker, the first step should be to check the batteries according to the instructions furnished with your detector.

Usually the method of testing the batteries will run something like this. Turn on the detector and adjust the meter for proper detecting before allowing it to stay on for five minutes. This will insure that you get an accurate battery test because many bad batteries will read "good" when they are first energized but will rapidly drop in voltage after they have warmed up.

After the batteries have warmed up, activate the battery test switch. If the battery is good the meter needle will indicate this on the instrument scale. Sometimes extremely fresh batteries will cause the needle to move off the scale, but this is normal. If the battery reads below normal replace it with a new battery. Repeat the operation on all batteries in the pack or section.

Remember also that just because a battery is new doesn't mean it can't be bad. The volume may read "good," but the capacity of the batteries to deliver current at the required low impedance or battery resistance may be poor. Partially discharged batteries can also read "good" when they are first turned on but may quickly discharge.

If you still have no speaker volume or sound after checking and replacing all defective batteries perhaps your instrument is tuned wrong. Carefully rotate the tuning control knob from one end of the band to the other until the detector is tuned. If you are sure that the instrument is properly tuned, check to see if the search coil lead is plugged in and making good contact.

Poor Sensitivity or Detection Capability

Again low battery voltage or high impedance batteries could be causing the problem. Check the batteries as just described and replace any that are defective.

The instrument may not be tuned correctly. If in doubt, reread the manufacturer's instructions carefully to make certain you are operating the detector correctly.

Also remember that the object on which you are testing the detector may not have been buried long enough to be detected as deeply as you think it should. An object buried for over a year will be detected more deeply than a fresh one. Try testing your instrument under actual field conditions and let the objects you normally dig be your indication of the instrument's sensitivity. Extremely dry soil and mineral interference will also cause a loss in depth penetration.

Excessive Frequency Shifting or Instability

The most common cause of this problem is low battery voltage or high battery impedance. Check and replace any defective batteries as already described.

Also, trying to use your instrument before it warms up can cause this problem. Always allow 10 to 15 minutes of warmup time before using your detector; this will allow it to reach its normal operating frequency at the prevailing temperature level. Batteries will also come to the "steady" voltage level during this warmup time.

If this does not solve the problem check for loose batteries or search coil connections; they should be tight.

Sometimes water leaks into the search coil and will cause frequency shifting. If you use your instrument on a hot summer's day and then plunge it into cold water the cooling effect can cause a vacuum inside the coil. Such a vacuum can pull water in through tiny holes in the coil cable where the cable enters the coil housing. These tiny holes become punched in the cable when the detector is used in briars, cactuses, sharp rocks, etc. If water does get into the search coil, dry the coil out at a temperature of about 100°F then test the coil. Once you are certain that all the moisture has been dried out, seal the area around where the cable enters the coil housing with epoxy glue.

Operating your instrument over highly mineralized ground can also be the cause of frequency shifting. So test your instruments in several locations before shipping it back to the factory.

Erratic Operation

This is another common problem that is usually caused by bad or weak batteries. Batteries always build up to higher voltage levels when they are shut off and when this high voltage is in the operating range of the detector the instrument will work perfectly for a few minutes after it is first turned on. However, this voltage drops rapidly and if it drops below the required operating voltage the detector responses can become erratic.

This problem could also be caused by loose battery terminal connections. Check and clean the battery terminals periodically and insure that all connections are secure. Check the coil leads for a tight connection and make certain that the leads are wrapped tightly around the detector stem. Loosely wrapped coil cables have been known to cause this problem in some types of metal detectors.

Erratic-Sounding Speaker

Small bits of dirt or other foreign materials sometimes drop through the speaker cover holes and cause the speaker to give off erratic sounds. To remove the particles, turn the detector upside down and shake it. If the pieces do not fall out, repeat the procedure while turning the instrument through its slower beats (if of the BFO type). The vibration of the speaker sometimes causes the objects to fall out. If the particles are iron, the speaker magnet may attract them sufficiently so that they do not fall out. In this case, hold a low power magnet near the speaker while the detector is held upside down with the instrument beating at a slow rate.

Note that in the above list most of the troubles are related to battery problems. This means that at least one extra set of batteries should always be carried by the treasure hunter when using his instrument in the field. One bad battery could ruin an otherwise enjoyable treasure-hunting outing.

You will also want to carry a tube of epoxy cement with you to secure any parts that may come loose. One of the most common items to break is the plastic flange that connects the aluminum search stem (shaft) to the search head. A tube of epoxy cement and some fine wire can help get you through the day until the instrument can be permanently repaired at home. It's even possible to carve a new shaft out of wood and secure it with epoxy and wire.

If none of the previously mentioned suggestions correct your circuitry problems, ship your detector back to the factory for repairs.

Here are a few troubleshooting tips for metal detectors that you have built yourself. Most problems occur in homemade detectors mainly because of poor connections and soldering. Therefore, many troubles can be eliminated by a careful inspection of connections to make sure they are soldered properly. Resolder any doubtful connections.

Check each circuit board to be sure there are no solder bridges between adjacent connections. To remove any unwanted connections, first hold the circuit board upside down. Place a clean soldering iron tip between the two points that are bridged until the excess solder flows down the tip of the soldering iron or gun.

Be sure each transistor is in the proper location and that each transistor lead is positioned properly and has a good solder connection to the foil.

Always check capacitor values carefully and make certain that the proper one is wired into the circuit at its proper location. Also check the polarity of electrolytic capacitors to be sure they are installed correctly.

Check each resistor carefully. It is very easy to get these mixed up—especially if you're an amateur starting on your first project. It would be easy, for example, to install a 1000Ω (brown-black-red) resistor where a 100K (brown-black-yellow) resistor should be. A resistor that is discolored or cracked or bulged is faulty and should be replaced.

If diodes are used in your detector circuitry, be sure the correct one is installed at its proper location. Also be certain that the banded end is positioned correctly.

Recheck the wiring by tracing each lead and comparing it to the schematic used to build the instrument.

Check all component leads connected to the circuit boards and make sure the leads do not extend through the circuit board and make contact with other connections or parts. You might want to test the circuits with an ohmmeter and a voltmeter once you are certain that all connections are properly made.

The finest metal detectors on the market today are those which have advanced circuitry and mechanical stability. Both are equally important. Therefore, you must make certain that all components and connections are firmly secured because poor mechanical stability will result in drift of the frequency. In general, rigid mounting of all components can be obtained by using short lead lengths on the parts and by using epoxy cement to hold any component that tends to shake or move in any way when the unit is moved. After your circuit board is completed, give it a good firm shake while observing the circuit components. If anything moves, dab several drops of epoxy on the guilty component and let it set until hard.

Some of the inexpensive factory kits could also stand some examination before using them in the field. Check these to make sure that the search coil lead is firmly wrapped around the search stem and that there is a tight connection on both ends—at the search coil and at the circuit board housing. Check all the connections on the circuit board to insure a good connection and to make certain that no bridging occurs.

A typical factory-prepared troubleshooting chart is shown in Fig. 14-1. Though this chart was designed for a Heathkit TR detector, it can also be used with other detectors. Of course, the part numbers may vary.

If you have access to test equipment, the use of it may save you some time and help you locate difficulties quicker. For example, a voltmeter may be used to check the voltages in your unit against the voltages shown in a schematic.

If you do use test equipment, be sure that you do not short any terminals to ground when making voltage measurements. If the

DIFFICULTY	POSSIBLE CAUSE
No sound in the speaker and no meter deflection when the unit is turned on and the Sensitivity control is advanced fully clockwise.	1. Battery dead. 2. Phone jack wired incorrectly. 3. Transistors or other components incorrectly installed. 4. Search coil incorrectly connected. 5. Pickup coil leads shorted. 6. Broken lead to battery. 7. Search coil leads shorting together.
Sound is heard from the speaker and meter indicates full scale regardless of Sensitivity setting.	1. NULL control R301 not adjusted for null. 2. Search coil or pickup coil improperly connected. 3. Search coil leads shorting together.
Unable to obtain a minimum reading during adjustments.	1. NULL control R301 incorrectly wired. 2. Search coil or pickup coil improperly connected. 3. Search coil leads shorting together.
Meter does not indicate or indicates backwards. Audio circuits operate properly.	1. Phone jack wired incorrectly. 2. Meter leads reversed. 3. Q207. 4. Meter 301. 5. Search coil leads shorting together.
Poor sensitivity.	1. Weak battery. 2. Improper adjustment (refer to the Operation section). 3. Transistors Q103, Q104, Q201, Q202, Q203, Q205, and/or Q206. 4. Search coil or pickun coil have shorted windings or leads. 5. Earth is so conductive that metal objects are not detected.
Erratic response when searching at a constant height above a small area of ground containing unknown objects.	1. NULL control adjusted so ferrous metals produce a decrease in the meter indication and tone, and nonferrous metals produce an increase. Refer to the "Operation" section of the Manual.

Fig. 14-1. A factory-prepared troubleshooting chart. (Courtesy Heathkit.)

probe should slip, for example, and short out bias or supply point, it will very likely cause damage to one or more of the transistors or diodes.

A transistor checker provides the most accurate check of transistors. However, an ohmmeter may be used to determine the general condition of a transistor, provided the ohmmeter used has at least 1V DC at its probes in order to exceed the threshold of the diode junctions inside the transistors. First, remove the transistor from the circuit and then set your ohmmeter on the R × 1000 range. Before performing any tests, however, make certain that you have properly identified the emitter, base, and the collector. Continue by connecting one of the ohmmeter test leads to the base lead of the transistor and then touch the other ohmmeter lead to the emitter and then to the collector. Both readings should be the same but may be either high or low. If one reading is high and the other low, the transistor is damaged and should be replaced. Repeat this last step with the ohmmeter test leads reversed. If all the readings are either low or high—no matter which ohmmeter lead is connected to the base—the transistor should be replaced.

Chapter 15

Hunting Valuable Plants

Many families participate in identifying, photographing, and even eating wild plants of all kinds. But did you know that many plants have value that can be readily converted into cash? Some medicinal plants have always been worth money and about the only equipment needed for hunting them is a digging tool and a container to carry your find in. You will, of course, need a knowledge of the plants you're looking for. And you'll need to know where to sell them once they are harvested. This chapter will acquaint you with the various wild plants that are the most popular and valuable. It will also tell you where to sell them for the most money.

GOLDENSEAL

Goldenseal is an herb (Fig. 15-1). Both its roots and leaves are used to make medicines. It is also known as yellowroot, yellow puccoon, orange-root, yellow Indian-paint, turmeric-root, Indian turmeric, Ohio curcuma, ground raspberry, eye-root, eye-balm, yellow-eye, jaundice-root, and Indian-dye.

This plant may be found in patches growing in high open woods on hillsides affording natural drainage. Its range is from southern New York to Minnesota and western Ontario, south to Georgia and Missouri. It is especially plentiful in the Allegheny Mountains. In other areas, it is becoming very scarce, enough so that a find is, indeed, a treasure.

Fig. 15-1. Goldenseal.

Goldenseal is a perennial plant belonging to the buttercup or crowfoot family. It has a thick yellow rootstock which sends up an erect hairy stem about 10 to 14 inches in height. The stem's base is surrounded by two or three yellowish scales. The stem is usually purple and has only two leaves, one large leaf and a smaller one. Sometimes, however, a third leaf, which is smaller than the other two, is produced.

The fresh rootstock of the goldenseal has a rank, nauseating odor and is bright yellow on both the inside and outside. It is one and a half to two and a half inches in length and one quarter to three quarters of an inch thick. It contains a large amount of yellow juice.

The stems and leaves (or herb) should be collected in midsummer. The roots should be collected in autumn after the seeds have ripened. They should be cleaned and then carefully dried in a shady place. The current price of goldenseal runs from about three to five dollars per pound for the herb and up to $12 per pound for the dried root. It may be sold to any of the following dealers:

William J. Boehner & Co., Inc.
259 W. 30th Street
New York, New York

Black Brothers, Inc.
141 Second Avenue So.
Nashville, Tennessee 37201

O.C. Plott
4058–4062B Peachtree Road NE
Atlanta, Georgia 30319

Frank LeMaster
Londonderry, Ohio 45647

Missouri Root & Herb Co.
Box 544
Hannibal, Missouri 63401

SNAKEROOT

This is another wild plant that is sought by root collectors for its medicinal value. Other common names include asarum, wild ginger, Indian ginger, Vermont snakeroot, heart-snakeroot, southern snakeroot, black snakeroot, and colt's foot. This inconspicuous little plant frequents rich woods or rich soil along roadsides and river banks from Canada south to North Carolina and west to Kansas.

It is a small, apparently stemless, perennial plant not more than six to 12 inches in height. It belongs to the birthwort family. It usually has two kidney-shaped leaves which develop from slender, hairy stems. The root has a creeping, yellowish rootstock, slightly jointed, with rootlets produced from joints which occur about every half inch or so.

The aromatic root of this plant is collected in autumn and the going price is normally low—around 15¢ per pound of dried roots.

SERPENTARIA

This is a much sought plant which brings a good price from the drug trade. It is found in rich woods from Connecticut to Michigan and southward along the Alleghenies. About midsummer the queerly shaped flowers of this native perennial are produced and are similar to those produced by the better known Dutchman's Pipe. Virginia serpentaria is nearly erect, the slender wavy stem sparingly branched near the base, and usually grows to about a foot in height, sometimes reaching a height of three feet. The leaves are thin and are either ovate, ovate lance shaped, or oblong lance shaped. The leaves are usually heart shaped at the base; they are about two and a half inches long and about one and a half inches wide. The flowers are produced from near the base of the plant, similar to its relative the Canada snakeroot. The flowers are solitary and terminal, born on slender, scaly branches. The fruit is a roundish six-celled capsule, about a half inch in diameter and containing numerous seeds.

The root is short with many thin, branching, fibrous roots. In the dried state the root is thin and bent, the short remains of stems showing on the upper surface and the under surface having numerous thin roots about four inches in length—all of a dull yellowish brown color on the outside and white on the inside. The dried roots bring five dollars a pound and up.

SENECA SNAKEROOT

This valuable plant is also known as senaga snakeroot, seneca-root, rattlesnake-root, and mountain flax (Fig. 15-2). It grows in rocky woods along hillsides and is found from New Brunswick and western New England to Minnesota and the Canadian Rocky Mountains, south along the Allegheny Mountains to North Carolina and Missouri.

The perennial root of this little plant sends up a number of smooth, slender, erect stems, sometimes slightly tinged with red, and generally unbranched. The leaves alternate on the stem and are lance shaped or oblong lance shaped, thin in texture, one to two inches long, and stemless. The flowering spikes are crowded, small, and greenish white. The flowering period of Seneca snakeroot is

Fig. 15-2. Seneca snakeroot.

from May to June in most areas. The plant is a member of the milkwort family.

The root is usually cylindrical and tapering; it is three to 15 centimeters long and 2 to 8 millimeters thick and bears several horizontal branches and a few rootlets. The time for collecting this root is in autumn. The root brings some of the better prices in the root collecting trade—somewhere around that of the goldenseal. Buyers of goldenseal will also buy seneca snakeroot.

BLOODROOT

Bloodroot (Fig. 15-3) is found in rich, open woods from Canada south to Florida and west to Arkansas and Nebraska. Though this plant's roots bring only about 50¢ per pound of dried roots, it is fairly plentiful and a large quantity may be gathered in a short time.

This indigenous plant is among the earliest of our spring flowers. The waxy-white blossom (Fig. 15-4) enfolded by the grayish-green leaf usually makes its appearance early in April. The stem and root contain a bloodred juice from which the plant got its name. Bloodroot is a perennial and belongs to the same family as the opium poppy. Each bud on the thick, horizontal rootstock produces but a single leaf and a flowering scape reaching about six inches in height. The plant is smooth, and both stem and leaves, especially when young, present a grayish green appearance, being covered with a "bloom" such as is found on some fruits. Each leaf has five to nine lobes; the lobes are either cleft at the apex or have a wavy edge. The leaves have leaf stems about six to 14 inches long. The leaves expand until they are about four to seven inches long and six to 12 inches broad. The under side of each leaf is paler than the upper side and shows prominent veins.

The mature root is thick, round, and fleshy, slightly curved at the ends, containing bloodred juice. It is one to four inches long, one half to one inch thick, and reddish brown on the outside.

Upon drying, the root shrinks considerably and the outside turns a dark brown while the inside is orange red or yellowish with numerous small red dots. It has a slight odor and is poisonous.

Fig. 15-3. The bloodroot plant. The dried roots of this plant bring over 50¢ per pound.

Fig. 15-4. The flowers of the bloodroot plant.

The rootstock should be collected in autumn after the plant's leaves have died. And after curing it should be stored in a dry place because it rapidly deteriorates if allowed to become moist.

GINSENG

Of all the medicinal roots collected for selling, ginseng is still the leader. At this writing, a pound of dried ginseng roots is worth about $65.

The earliest Chinese book on the medicinal value of herbs was written by Emperor Shen-ung around 3000 B.C. In it, ginseng is regarded as the most potent among thousands of herbs mentioned. The plant's name was composed from two Chinese characters meaning "manplant" or the "shape of man." Indeed, quite a few of the roots bear a remarkable resemblance to the body of a man.

With the exception of tea, ginseng is probably the most celebrated plant in the Orient. Many Chinese believe it a cure for fevers and illness of both body and mind. It is a very valuable folk medicine in today's China.

The natural range of ginseng, growing wild in the United States, extends all the way from the Canadian border to Alabama. But the American variety of ginseng was not considered of much medicinal value until 1905 (except by the Chinese who had been using it for medical purposes for centuries).

Many American authorities on the subject still have doubts as to the root's medicinal value. However, there seems to be no doubt in the minds of many Chinese since there are still single orders from China for as much as 1000 pounds! It seems that the roots from the northern states are the most sought after variety.

Ginseng sprouts first appear around the first of May (Fig. 15-5). At two years of age, the plants have several light green leaves (Fig. 15-6). When at its full growth it has three leaf clusters branching from one stalk and each leaf stem has five leaves (Fig. 15-7). It

Fig. 15-5. The ginseng plant in early spring.

Fig. 15-6. The two-year-old ginseng.

Fig. 15-7. The mature ginseng.

reaches the height of two feet. The stalk dies down every fall (Fig. 15-8), and near the neck of the root, a scar remains to tell the age of the plant.

Though ginseng has been successfully cultivated in recent years, the Chinese still prefer the wild plant. This is where treasure hunters come in.

The American ginseng can hardly be distinguished from the Chinese plant. Nevertheless, the ginseng has several different grades:

First Grade—raised in China's former dynastic land holdings
Second Grade—raised in other areas of China
Third Grade—raised in America
Fourth Grade—raised in Japan and Korea

The home of the wild ginseng in this country is in forests where it is shielded from the direct rays of the sun. It is rarely found in

Fig. 15-8. The leaves of the ginseng plant turn yelllow in the fall before dropping off.

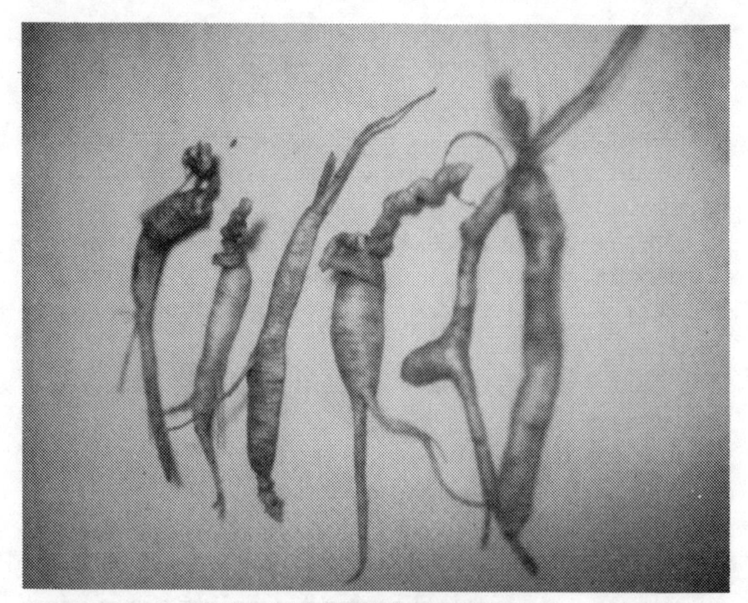

Fig. 15-9. Fresh ginseng roots.

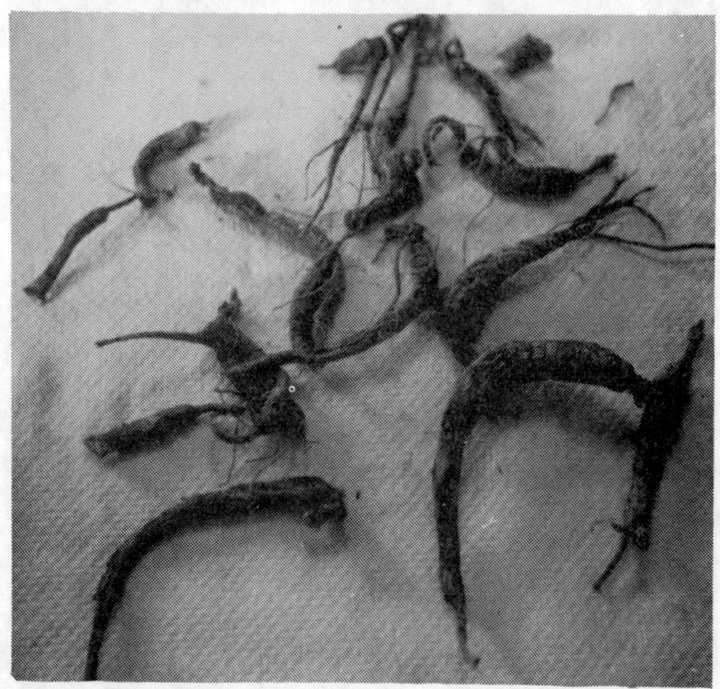

Fig. 15-10. Ginseng roots after they have dried for about two weeks.

damp, swampy, or sandy places. The four varieties of the plant in the United States are distinguished by the following names: northern, southern, western, and round.

Many people hunt this root, but the experienced hunter doesn't waste time searching in unprofitable places. He understands its nature and environmental needs. Once the correct environment is found—ginseng is usually found. Remember that moisture is its greatest enemy; the second greatest is heat.

Ginseng roots are usually dug in the autumn and great care must be taken so they won't become damaged. They are first washed in water, using a soft brush. Then they are laid out in a dry place away from the direct rays of the sun (Fig. 15-9). This is important because much of the value of the root depends on the way it is prepared. After the roots have dried for about two weeks, they will look like the ones shown in Fig. 15-10.

Even with its high price, don't count on getting rich from digging ginseng. Many diggers around the Virginia area seldom find more than four or five pounds in a season. Still, there is a chance that you could run into a virgin patch where a hundred pounds or more could be dug. This would bring you over $6000!

Chapter 16
Fossils

When searching for the more common treasures don't overlook the possibility of finding fossils—remnants of prehistoric life. Many treasure hunters, especially skin divers, have come across bones and fossils. The value of a fossil depends entirely on the fossil itself. If it's common, the value will not be too great. If your find is rare, it could be worth a great deal.

Very few plants and animals that died in prehistory are preserved as fossils. This is because in order to become fossilized, the plant or animal had to become buried quickly—very quickly—to prevent decay and it then had to be left undisturbed for thousands of years.

The question that often arises is where do you find fossils? Actually fossils are probably everywhere waiting to be found, but most of the fossils that are found are nearly always in sedimentary rocks. These rocks are formed by sediment—mud, clay, sand—deposited in seas, lakes, caves, deserts, or river valleys. Sedimentary rocks occur over much of North America. But in many places the fossil-bearing layers lie deep beneath other rocks. Therefore, fossil hunting, for most people, will be limited to places where the sedimentary rock is exposed at the surface—in cliffs, river banks, road cuts, or quarries. Places like these are where to begin looking.

Many major fossil discoveries have been made by amateurs, and many amateurs have won acclaim from professional geologists. Take, for example, a recent discovery on Thunderbird Ranch near Front Royal, Virginia. In the early 1970s a local land developer purchased the 1000-acre tract of land. One strip of this land, located along the Shenandoah River, was searched by an amateur archaeologist for Indian artifacts. Indian artifacts were found in the area and eventually the site proved to be one of the most important archaeological sites in North America! So far, artifacts found on the site cover a span of human development from 11,500 to 8,000 years ago. Fossils of mammoths, mastodons, and early species of horses and camels were also found. Caves are now under excavation with the hopes of uncovering other important traces of the past.

There are other lands near this property which are likely to yield other valuable finds and permission to search can be had from local landowners. The George Washington National Forest is close by, and it may yield some fossils and artifacts, too.

If you have ever been a rockhound, then you probably have most of the tools you'll need for fossil hunting; that is, a geologist's pick, a knapsack or shoulder bag for carrying specimens, newspaper for wrapping specimens, labels for specimens, a small cold chisel, a small shovel, a steel wrecking bar, road maps, topographic or geological maps, a magnifying glass, a compass, a notebook, a pencil, and a pocket knife.

When you have decided upon a probable area, take time to look the area over before starting to dig. Look for rock surfaces where weathering has exposed fossils. Your next step should be to turn over the smaller rocks and study all sides, breaking open concretions as you go. If you should locate bones or fossils you believe rare, leave them in intact. Get a professional to examine them for you.

Once specimens have been removed they will have to be cleaned and prepared to prevent deterioration. The delicate task of cleaning a specimen is best done on a stout table with good light and adequate tools. Old dental tools are excellent for this process as are small electric drills or grinders (the kind used by hobbyists). A coating of shellac should be applied to all bones and delicate fossils to prevent their cracking and deterioration.

Consult books on fossils for identification of your specimens. If you don't find what you're looking for in books, secure the aid of a geologist at the closest university or museum.

Fossils can be stored in cardboard trays purchased from scientific supply houses. Labels may be placed in the tray for identification purposes.

Reference materials are very important in fossil hunting. Without them, locating fossil deposits would be nearly impossible. Maps are essential reference tools; you should have three kinds:

Road Maps: Obtain several sets from your local service station. Keep one set at home for reference and take the other set into the field. Both sets should be kept up to date.

Topographical maps: These maps are produced by the U.S. Geological Survey in Washington, D.C., and are more detailed than conventional road maps. They show all land and water features, including roads, bridges, towns, houses, etc. An Index Map for your area can be obtained from the U.S. Geological Survey; then you can choose what maps you'll need for your particular activities.

Geological maps: These maps are prepared by state geological surveys, either as a state geological map or in relation to specific reports. Check with your state survey and get their list of publications.

Books are also invaluable reference aids to the fossil hunter. The selected bibliography in Appendix 3 lists books that cover all phases of fossil hunting and identification.

Chapter 17

Treasure Hunting For Tenderfeet

Almost all the previous data in this book has been about hunting treasures outdoors. This chapter, though, is designed to give you the basic knowledge to enable you to begin your nonoutdoor treasure hunting—at auctions, estate sales, and other places. Let's begin with a few of the most likely objects of value to be found locally—maybe right in your own attic!

AMERICAN PAINTINGS

Less than two decades ago, American art—with few exceptions—was generally not in demand. But many American paintings that sold for around $2000 some years ago have brought over $200,000 at recent auctions. And there is a good chance that many American paintings just as valuable are collecting dust in many attics throughout the United States.

The best place to obtain such paintings at a bargain is at an estate auction sale—one in which it is up to the heirs to sell everything and divide the proceeds equally among themselves. Look in the newspapers for auction sales of this type.

Become familiar with all famous artists and learn to recognize their paintings. You just might run into a real find at your next visit to an auction sale.

ENGLISH WATERCOLORS

English watercolors—those capturing pleasant landscapes or seascapes with muted colors—are other paintings which are "big game" for treasure hunters. An 8×10-inch painting by J.M.W. Turner, for example, may bring over $100,000. Again, look to the auctions for bargains on these.

In order to become familiar with English watercolors, purchase the book *English Water-Colours* by Stanley W. Fisher (Ward Lock Limited, London). This is a small, easy-to-read volume which includes plates of the works of many of the watercolorists of the 18th and 19th centuries. The information in this book should give you a good idea of just what these watercolors are, the prices being asked for them, and the trends of the market.

DRAWINGS

Good drawings done in pencil or charcoal can be very valuable. And the chances of finding one of these are good.

For example, this is how one treasure hunter made a find. In 1960, a sketch—in brown and black, heightened with white—appeared in the art shop of Herman Galka in New York City. The name of the sketch was "The Prodigal Son Among the Swine." Since Galka was not certain of the artist, he offered the drawing for $70. A customer noticed the sketch and said that he might be interested in purchasing it if he could take it on approval for study. The dealer agreed and the sketch was eventually purchased for the sum of $60. Later the drawing was shown to Robert L. Manning, director of the Finch College Museum of Art, who placed the drawing on display some time later. While it was being exhibited, another collector saw the drawing and recognized it as a work by Costiglione, the Italian baroque artist. After some negotiating, the collector purchased it for $2250. This meant almost $2200 profit for the original purchaser. Later on, this same painting was sold to David Rust of the National Gallery in Washington, D.C., for approximately $10,000.

OTHER TREASURES

Old paintings and drawings aren't the only treasures around that can bring big payoffs. There are sculptures, antiques, silverware, glassware, paperweights, porcelain, pottery, rugs, books, autographs, jewels, wines, old cars...the list is endless, and the key to finding all these things is research.

Look for reference or pricing books in your local library. Learn how to recognize the objects of value; then begin your treasure-hunting expedition by combing the antique shops, auction sales, old homes, etc. Your chances of finding a real treasure are probably as good as those of any treasure hunter with a metal detector.

Chapter 18
Coping With Some Hazards Of The Outdoors

Treasure hunting in the outdoors is far safer—in most cases—than driving on the freeway or crossing a busy street. Yet anyone who spends a great deal of time treasure hunting in the outdoors is eventually going to encounter some of nature's hazards—like snakes, poison ivy, etc. Avoiding these hazards just takes a little knowledge and a little forethought.

Sooner or later you're going to run into poison ivy. It's found in practically every state in this country and thrives in any type of soil that contains sufficient moisture. The plant is easily recognized by its leaves which always grow in clusters of three (Fig. 18-1). Poison ivy plants can entwine themselves around everything from trees to fences to old buildings. And if they can't find anything to climb on, they merely scramble along the ground.

Many people think they are immune to poison ivy. It is true that some individuals coming in contact with it for the first time do have some degree of resistance to the plant's oily poison. But continuous contact with the plant will gradually increase sensitivity until the individual loses all resistance.

It is almost impossible for someone who does a lot of outdoor treasure hunting to avoid contact with poison ivy. So if you're going to be searching through woods or fields and the like, wear proper clothing to prevent becoming infected. Put on a long-sleeve shirt, trousers, high-top shoes or boots, and gloves. If you're going to be digging in the ground for buried treasure, you will probably be wearing gloves anyway. When you return home, remove these clothes immediately and take a bath, using plenty of soap. Then wash the clothes you were wearing; don't put them on again until you have done this. If you follow these procedures religiously, the chances of your getting an infection are nearly nil.

If you cook over campfires while on your outings, avoid burning logs that have poison ivy leaves on them. Even when the poison ivy vines have been dead for several weeks, they are still capable of causing infection. The heat from the fire will break down the oil into very fine droplets, which are then automatically carried by the smoke. And when an individual inhales the smoke...possible infection on the lips, in the throat, and even in the lungs. So be extremely cautious about this practice.

If you do get a poison ivy infection, the use of Aristocrat 0.16% Cream will check it immediately. This cream is a cortisone-base cream requiring a doctor's prescription. When applied to the infected areas as soon as the first signs appear then reapplied every hour or so thereafter, the itching will stop and the infection will be completely dried up in a day or two.

For severe cases, you can attack the infection internally as well as externally. That is, you can apply the cream directly to all infected areas and then take cortisone pills internally (as prescribed by a physician). This drug lessens the agony and usually ends the infection within four days—even in severe cases.

Animals are not allergic to poison ivy. So you don't have to worry about your pets becoming infected. They can, however, carry the poison ivy oil on their fur and pass along the infection to you!

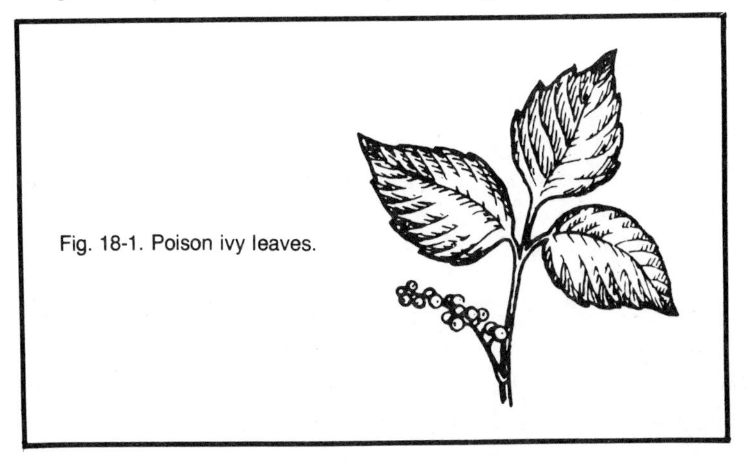

Fig. 18-1. Poison ivy leaves.

Snakebite in the wilds is very rare, indeed. When it does occur, it's usually because someone has been careless about observing the following rules:

1. Never step where you can't actually see and never place your hands over or around an object when you cannot see the surface (or any small niches) your hands will touch or come near.
2. Always wear boots, or at least high-topped shoes, when wandering into snake territory.
3. When camping, always shake out your shoes and clothing before dressing.
4. Always be on the alert for any movement on the ground in front of you as you are walking in snake country.
5. Never play with a snake, whether you think it's poisonous or not. Even nonpoisonous snakes can cause infection.

If you should get bitten by a poisonous snake, place ice packs on the bite and get to a doctor. The ice will relieve some of the pain, keep the swelling down and perhaps retard the rate at which the poison enters the skin tissue. If you have a snakebite kit handy, use it.

Be careful about drinking the water you find in the outdoors. It's not all safe for consumption. If you should have to use questionable water in an emergency, purify it first by adding a water purifier (chlorine, etc.) or by boiling it for at least 20 minutes.

Appendix 1

Some Treasure And Hunting Terms

all-purpose—A term used to describe some of the better metal detectors on the market; they are designed for the full range of treasure-hunting capabilities.

BFO—Abbreviation for beat frequency oscillator, a type of metal detector that indicates a find by a change in audible pitch.

coaxial search coil—A stacked loop type of search coil used on VLF metal detectors; it gives uniform response and fringe-area detection. It also prevents interference from 60-cycle power lines and detectors operating nearby.

coin shooting—Hunting coins with a metal detector. A person who undertakes this hobby is known as a "coin shooter."

discriminator—A metal detector which will, by means of electronic circuits and meter response, indicate the difference between wanted and unwanted finds.

ditty—A small treasure cache.

double coils—Two search coils or loops operating independently on the same metal detector.

drift—An irritating metal detector tuning problem caused by varying weather conditions, ground reactions, etc.

float—Rich ore that has fallen or washed away from the lode.

hot—Very sensitive (pertaining to the performance of metal detectors). Negative and positive soils are referred to as "hot soils."

narrow scan—A term used to describe most TR search coil scanning widths.

negative soil—Ground in which iron content makes a metal detector's tone or frequency decrease as the detector coil is moved toward the ground.

neutral soil—Ground which is the best for scanning with a metal detector.

placer deposits—Concentrations of tiny pieces of gold, silver, or other minerals which are found in the fissures of creek beds.

positive soil—Ground in which salts and other concentrations make a metal detector's tone or frequency increase as the detector coil is moved toward the ground.

real—A silver coin of a former Spanish monetary system. Eight of them make a dollar; thus, they are called "pieces of eight."

reis—The plural of *real*.

rockhound—An amateur rock and mineral collector.

sand—Gold dust or flour.

score—A successful treasure hunt or find.

sponge—A porous piece of gold or gold alloy, usually produced by retorting the mercury from a gold-mercury amalgam.

stability—The ability of a metal detector circuit to function without drift.

THer—Treasure hunter.

THing—Treasure hunting.

TR—A transmitter-receiver metal detector (also called inductive balance detector); such a detector indicates a find by a change in signal intensity.

trove—A discovery or find.

VLF—A very low frequency metal detector. Such a detector generally operates in the frequency range of 500 to 2000 Hz per second. This type of detector is also known as a(n) GEB, MFO, Magnum, and GCD.

waybill—A map or instructions to a secret map or treasure cache.

wide scan—A recently developed TR search coil with wide scanning capabilities.

Zero-Drift—A patented development which eliminates all troublesome drift in a metal detector's operation.

Appendix 2
Equipment Dealers

The following dealers specialize in metal detectors and other treasure-hunting equipment.

Alabama

G & B Detector Sales
BMK Building
1378 Highway 31 S.
Alabaster
(205) 663-4573

Birmingham Coin & Relic
Detector Sales
113 24th Ave. NW
Birmingham 35125
(205) 854-0341 or
854-3364

Dan Owens
131 S. 11th St.
Gadsden 35901
(205) 546-6561

Alaska

Stewart's Photo Shop
531 Fourth Ave.
Anchorage 99501

Arizona

McCowan Enterprises
824 E. Broadway Rd.
Phoenix 85040
(602) 276-6774

Arkansas

W. W. Mosely
P.O. Box 7
Camden 71701
(501) 836-5314

Herring Electric Co.
1217 W. Hillsboro St.
Eldorado 71730
(501) 862-3928

Bill's Detectors
5623 "R" St.
Little Rock 72207
(501) 666-6355

L. L. Lincoln
Rt. 1
Rogers 71756
(501) 636-6867

California

General Electronic DTCTN
16238 Lakewood Blvd.
Bellflower 90706
(213) 633-5330

Aurora Prospectors Supply
6286 Beach Blvd
Buena Park 90620

Rencher Welding & Machine Works
560 S. Third St.
Chowchilla 93610
(209) 665-4219

Escondido Coin Shop
111 N. Broadway
Escondido 92025
(714) 745-1613

Gene Rolls
Highway 32 at Sugar Pine
Forest Ranch 95942
(916) 342-4829

Fresno Hobby & Crafts
3026 N. Cedar
Fresno 93703
(209) 226-4880

Central Cal Electronics
600 East St.
Hollister 95023
(408) 637-1704

The We Shop
221 Main St.
Huntington Beach 92648
(714) 536-4700

South Bay Coins
639 Ninth St.
Imperial Beach 92032
(714) 423-2551

Fumble Fingers
3384 Mt. Diablo Blvd.
Lafayette 94549
(415) 284-7406

White's Detectors
Arthur Apodaca
3141 E. Highway 140
Merced 95340
(209) 723-0636

Gold Nugget Miners Supply
1302 Ninth St.
Modesto 95354

Gems Galore
240 Castro
Mountain View 94040
(415) 968-8707

The Coin Shop
1516 Third St.
Napa 94558
(707) 255-8166

Treasures Unlimited
526 66th Ave.
Oakland 94621
(415) 638-6352

N & A Belle Electronics
126 W. Holt Blvd.
Ontario 91761
(714) 984-9119

Buddy Sales
18552 Sherman Way
Resenda 91335
(213) 342-5113

Pioneer Recoveries
3510 Audubon Place
Riverside 92501
(714) 682-4302

Mother Lode Diving Shop
2001 Capitol Ave.
Sacramento 95814
(916) 443-3237

Coin Chest Coin Shop
545 West Baseline
San Bernardino 92410
(714) 885-5711

Gem & THing Association
2493 San Diego Ave.
San Diego 92110
(714) 297-2672

San Diego Coin Exchange
3784 30th St.
San Diego 92104
(714) 296-3131

Arts & Hobbies
12334 Kagel Canyon Rd.
San Fernando 91342
(213) 896-9477

Mining & Lapidary
131 10th St.
San Francisco 94101
(415) 626-6016

Dunall's Outdoor Supply
605-7 State St.
Santa Barbara 93191
(805) 963-5319

Hans Auto Supply
555 Charles St.
Seaside 93955
(408) 394-6044

The Treasure Cave
18 W. Bradford St.
Sonora 95370
(209) 532-5836

Wrights
2405 N. El Dorado St.
Stockton 95204
(209) 466-4351

Colorado

Suack Enterprises
5190 N. Nevada
Colorado Springs 80907
(303) 598-0701

The Prospector's Pick
#2 Ruxton Ave.
Manitou Springs 80829
(303) 685-1457

Dan & Reg's Treasure Instruments
P.O. Box 1697
Pueblo 81003
(303) 566-0698

Metal Detectors Sales
6117 W. 38th Ave.
Wheat Ridge 80033
(303) 522-4566

Connecticut

Edward Perchaluk
304 Circle Dr.
Stratford 06497

Florida

Underwater Unlimited
216 Palermo Ave.
Coral Gables 33134
(305) 445-7837

Harry's Pawn Shop
519 Main St.
Jacksonville 32202
(904) 353-6971

Edgewood Coin Shop
934 S. Edgewood Ave.
Jacksonville 32205
(904) 389-0013

Palm Plaza Camera
713 N. 15th St.
Leesburg 32748
(904) 787-4661

Kellyco
1811 Chinook Trail
Maitland 32751
(305) 642-2285

Tamiami Gun Distributors
2975 S. W. 18th St.
Miami 33135
(305) 642-1941

O Kenin, Inc.
1834 N.E. 163rd St.
N. Miami Beach 33162
(305) 949-7681

Wen-Co Enterprises
6874 80th Ave. N.
Pinellas Park 33565
(813) 546-9264

Dan Byrd Fish Camp
At New Pass Bridge
Sarasota 33577
(813) 888-2488

Treasure Shack
3934 Britton Plaza
Tampa 33611
(813) 833-9841

Georgia

S & H Metal Detector Sales
2052 Upper Ridge Rd. S.E.
Dalton 30720
(404) 226-3981

Finders Co.
225 Upland Rd.
Decatur 30030
(404) 377-0974

Ernest M. Andrews
2755 Sylvan Rd.
East Point 30344
(404) 777-8141

The Stamp Shed
1115-D Watson Blvd.
Warner Robins 31093
(912) 923-6159

J.C. Ballentine
Hatcher Point Mall
Waycross 31501

Idaho

Mountaineering Outfitter
62 N. Main, Box 116
Driggs 83422
(208) 354-2222

Outdoor Hobby Supply
2416 E. Main
Lewiston 83501
(208) 743-1768

Roy Norstrom
339 Chateau Thierry
Box 419
Soda Spring 84376
(208) 547-4166

Illinois

Rene's Treasure Trove
1101 N. Livingston
Bloomington 61701

Wheat State Electronics
12203 Vincennes Ave., Apt. 24
Blue Island 60406
(312) 371-7086

Jerry's Gun Shop
9220 Ogden Ave.
Brookfield 60513
(312) 485-5200

Gene M. Blaker
2903 W. John
Champaign 61820
(217) 356-0918

Harry's Treasure Shack
322 W. State St.
Cherry Valley 61016
(815) 332-5157

Frontier General Store
1066 W. Harrison
Decatur 62526
(217) 877-9199

Automatic Engineering
1630 Ogden Ave.
Downers Grove 60515

B-T Electronics
317 S. McLean Blvd.
Elgin 60120
(312) 742-3090

White's Metal Detectors
302 Poplar St.
Harrisburg 62946
(619) 253-5131

The Treasure Trove
727 S. Finley Ave.
Lombard 60148
(312) 629-3980

N & D Detector Sales
2110 N. Richmond Rd.
McHenry 60050

Dal-Mar Electronics
5815 W. Warren St.
Morton Grove 60053
(312) 966-9687

Rev. John J. Costas
3116 11th Ave. "A"
Moline 61265
(309) 797-3098

Pioneer Treasure Detectors
P.O. Box 55
123 Dee St.
Moro 62067
(618) 377-8655

Orland Treasure Finders
14231 S. Union Ave.
Orland Park 60462
(312) 349-2078

Dee's Beauty Shop
206 Reservoir Rd.
Pekin 61554
(309) 346-4377

B & E Sales
1009 S. Tonti Circle
Peoria 61605
(309) 637-1570

Mid-West Treasure Detectors
505 S. Eight St.
Quincy 63301
(217) 223-4723

Tom's Pool Center
809 N. Green Bay Rd.
Waukegan 60085

Indiana

O-D Western Store
R.R. #5
Decatur 46733
(219) 724-2097

Krull's Hobby Shop
414 E. Washington Blvd.
Fort Wayne 46802
(219) 422-4429

C & E Treasure Seekers
16340 Fairfield Ln.
Granger 46530

J & J Coins
7019 Calumet Ave.
Hammond 46324
(219) 932-5818

Hansen's Sports
Highland 46322
(219) 838-7495

Metal Locating Equipment
2916 W. 62nd St.
Indianapolis 46268
(317) 251-8472

Pollard Coin & Stamp
Supply Co., Inc.
5220 E. 23rd St.
Indianapolis 46218
(317) 547-1306

Iowa

Richard Cross
Box 90
Baxter 50028
(515) 227-3391

George & Dorothy Morgan
2003 Walnut St.
Cedar Falls 50613
(319) 266-0217

Geode Industries, Inc.
106-108 W. Main St.
New London 52645
(319) 367-2286

Spragg Electronic Service
1501 W. 13th St.
Davenport 52804
(319) 323-7444

The Treasure Chest
1525 S. Concord
Davenport 52804
(319) 322-9355

Kansas

Carl Clare
911 3rd Ave.
Dodge City 67801
(316) 225-4701

Bob Ruth
Everest 66424
(913) 548-3385

Epp's Coin Shop
112 S. Main
Pratt 67124
(316) 672-6181

Bonta's Treasure Shack
617 Chase
Wichita 67213
(316) 942-6965

Mid-Western Research & Supply
6323 Scottsville
Wichita 67219
(316) 744-0668

Kentucky

Lewis Eddleman
3320A Wood Valley Ct.
Lexington 40502
(606) 266-0980

A.F. Waller
10800 Marcitis Rd.
Louisville 40272
(502) 937-8998

Cecil & Red
2529 W. Market St.
Louisville 40212
(502) 778-2119

Louisiana

Clark's Detector Sales
3233 Illinois Ave.
Kenner 70062
(504) 443-1285

Woody's Coin Shop
520 Miller Ave.
Westlake 70669
(318) 439-9616

Maryland

Bay Side Treasure Detectors
422-430 E. 42nd St.
Baltimore 21218
(301) 235-4412

South Mountain Sports
Rt. #2 Box 344A
Keadle Rd.
Boonsboro 21713
(301) 739-5384

Frank White's Detectors
408 Arbor Dr.
Glen Burnie 21061
(301) 768-3157

Martins Metal Detector Sales
601 Robinhood Rd.
Havre De Grace 21078
(301) 939-1417

Albert M. Schone
430 Leyton Rd.
Reisterstown 21136
(301)833-6124

Robert F. Walters
Love Point Rd.
Stevensville 21666
(301) 632-6397

Massachusetts

E & D Electronic Sales
83 Parker St.
Agawam 01001

Ronimac Firearms Co.
530 Chaplin St.
Ludlow 01056
(431) 589-9096

Larry Violette
Box 74
Rehoboth 02769
(617) 252-4497

Michigan

Johnston's TV
Box 161
Baraga 49908
(906) 353-6858

Wayside Sports Center
1720 W. Brady Rd.
Chesaning 48616
(517) 845-3030

Detectors
5612 Croissant
Dearborn Heights 48122
(313) 274-4186

Grant's Book Store
601 Bridge St. N.W.
Grand Rapids 49504
(616) 458-6480

Treasure Hunter's Supply
1949 Collins SE
Grand Rapids 49507
(616) 243-0216

Electronic Metal Detectors
525 Riley St.
Holland 49423
(616) 396-5692

Electronic Center
1010 Washington St.
Marquette 49855
(906) 228-9670

Marshall Artifact Recoveries
205 W. Michigan Ave.
Marshall 49068
(616) 781-4575

Treasure Land
38554 Groesbeck Hwy.
Mt. Clemens 48043
(313) 469-0600

Minnesota

Treasure Sport Specialty Co.
1208 Park St.
Anoka 55303
(612) 421-8971

Mid West Detectors
11120 Vincent Ave. S.
Bloomington 55431

Schilling Coins
104 E. Superior St.
Duluth
(218) 722-8065

Sage Christian Enterprizes
321 W. First St.
Park Rapids 76470
(218) 732-5907

Minnesota Prospector's Supply
RR #4
Red Wing 55066
(612) 923-4728

Betlach Jewelers
8432 Hwy. 7
Knollwood Plaza
St. Louis Park 55426
(612) 935-4308

Herman Hafstad
Rt. 1
Willmar 56001
(612) 235-5224

Mississippi

Mid-South Treasure Barn
4226 Highway 80 W.
Jackson 39209
(601) 922-1663

Hobbies Unlimited
1219 Nelle St.
Tupelo 38801
(601) 842-6031

Missouri

R & R Stamp & Coin
Flat River 63601

Bob's Treasure Shack
8108 E. 80 Ten
Kansas City 64138
(816) 358-4928

Clevenger's Hobby
8206 North Oak
Kansas City 64118
(816) 436-0697

Ray George Company
3456 S. Grand Ave.
St. Louis 63118
(314) 776-4568

Paul Snyder
202 Owens
Smithville 64089
(816) 873-3339

Radford Jewelers
1864 S. Glenstone
Springfield 65804
(417) 881-7309

Frank Sales & Service
711 S. Jefferson
Webb City 64870
(417) 673-2650

Montana

J & L Coinfinders
2102 Miles Ave.
Billings 59103
(406) 656-6569

Mid-West Welding & Machine Shop
Box 1115
2320 N. Seventh Ave.
Bozeman 59715
(406) 487-5417

Farmer's Electric Supply
1998 20th St. S.
Great Falls 59401
(406) 453-5501

Electronic Parts
1030 S. Ave. W.
Missoula 59801
(406) 543-3119

H&B Welding Service
808 Third Ave.
Roundup 59072
(406) 323-2532

Nebraska

The Spartan Shop
335 N. Williams
Fremont 68025
(402) 721-9438

Canfield's
2415 Cuming
Omaha 68131
(402) 342-1517

L.P. Enterprises
Box 46
Sprague 68438
(402) 794-5730

Nevada

Karl's CB Sales
1201 Stewart
Las Vegas 89101
(702) 382-5011

Nevada Coin Mart Supply
2409 Las Blvd. S.
Las Vegas 89105
(702) 732-1375

New Hampshire

Don Wilson Sales
83 S. State St.
Concord 03301
(603) 224-5090

George F. Streeter & Co.
265 Washington St.
Keene 03431
(603) 357-0229

New Jersey

General Sales Co.
10 Humphrey St.
Englewood 07631
(201) 568-5563

Geo-Quest
78 Kenzel Ave.
Nutley 07110

Sherando Corp.
61 W. Kings Hwy.
Mt. Ephraim 08059
(609) 931-1113

The Treasure Cove
1055 S. Clinton Ave.
Trenton 08611

The Treasure Chest
Mt. View Electric Co.
866 Rt. 23
Wayne 07470
(201) 694-0777

New Mexico

Raydon Electronics
820 Sullivan
Farmington 87401
(505) 325-4883

New York

Tools Unlimited
2106 Union Blvd.
Bay Shore 11706
(516) 581-4882

Bigford Brothers
44 Clyde St.
Earlville 13332
(315) 691-3563

Panna's Electronic Sales
P.O. Box 167
Geneva 14456
(315) 789-0809

Bond Electric Co.
2 Leich Cir. Co.
Glen Cove 11542
(516) 676-1380

Trade Mart Enterprises
94 Keller Ave.
Kenmore 14216
(716) 875-0951

Wheels Afield Travel Trailer
Rd. #3 Box 142
Kingston 12401
(914) 331-5687

Harbor Coins & Stamps
134 Mamaroneck Ave.
Mamaroneck 10543
(914) 698-5671

Kay's Detector Sales
Rt. 9 Malta
Rd #1 M
Mechanicville 12128
(518) 899-4961

C-T Detectors
293 W. First St.
Mt. Vernon 10550
(914) 668-0518

M. Friedlander Hardware, Inc.
146 W. 54th St.
New York 10019
(212) 247-7774

Detection Enterprises of Rochester
366 Brookview Dr.
Rochester 14617
(716) 338-2588

Doc Dave's TV Service
54 Stockton Ave.
Walton 13856
(607) 865-5188

North Carolina

Pete's Electric
P.O. Box 3
Dickinson 58601
(701) 225-5025

Detector Sales Co.
131 Stetson Dr.
Charlotte 28213
(704) 596-1953

Science Hobbies, Inc.
2615 Central Ave.
Charlotte 28205
(704) 375-7684

Carson's Coin Shop
112 N. Center St.
Statesville 28677
(704) 872-9671

Ohio

Bennett Detector Sales
203 E. Market St.
Akron 44308
(216) 367-9664

Cutcher's Brownhelm Store
1605 N. Ridge Rd.
Amherst 44001
(216) 988-4110

Red's Sport
Rt. 39 & 62
Berlin 44610
(316) 893-2545

Aluebelhart's Sport Shop
2819 Fifth St. W.
Canton 44708
(216) 455-7837

Kilian Detector Equipment
1031 Spring Rd.
Cleveland 44109
(216) 398-4779

Harbor Coin
215 Third St.
Fairport Harbor 43027
(216) 354-2458

3-E Rock N Treasure Ctr.
638 Chestnut Ridge Rd.
Hubbard 44425
(216) 534-4482

Klingler's Rocks N Things
411 Bowman Rd.
Lima 45804
(419) 227-5294

Knapp's Metal Detector Sales
18944 Mitchell Ave.
Rocky River 44116
(216) 331-1658

Purcell's
109 S. First St.
Tipp City 45071
(513) 667-2143

Allen's Coin Shops
34 N. State St.
Westerville 43081
(614) 882-3937
and
1243 W. Broad St.
Columbus 43216

Oklahoma

B & J Treasure Shack
1219 WFP Blvd.
Bartlesville 74003
(918) 336-2155

Russey's Electric
620 Stadium Dr.
Hobart 73651
(405) 726-3686

Arrowhead Supply
330 S.W. 28th St.
Oklahoma City 73109
(405) 634-7128

Hobby World
2433 Plaza Prom
Shepherd Mall
Oklahoma City 73107
(405) 942-5556

Mid-Del TV & Electronics
4518 S.E. 29th
Oklahoma City 73115
(405) 677-1194

H & H Metal Detector Sales
114 S. Wewoka
Wewoka 74884
(405) 357-3137

Oregon

Carla Kay Salvage
471 Bruel St.
Coos Bay 97420
(503) 888-4015

Oregon Sporting Goods
175 Main
Hermiston 97838

Bennington Detector Sales
5059 Bryant Ave.
Klamath Falls 97601

Hal's Car Rental
903 Adams Ave.
La Grande 97850
(503) 963-3884

Pendleton Stamp & Coin
432 S. Main
Pendleton
(503) 276-6870

D & K Detector Sales
13809 E. Division
Portland 97236
(503) 761-1521

Young's Sporting Goods
515 E. Second
The Dalles 97058
(503) 296-2544

Pennsylvania

Bob's Treasure Shack
335 E 11th St.
Berwick 18603
(717) 759-1783

Milloway's Motorcycle
and Parts Shop
439 E. Eight St.
Berwick 18603
(717) 759-9372

J & D Metal Locating Equip.
Conemaugh 15909
(814) 749-9411

Gettysburg Electronics
27 Chambersburg St.
Gettysburg 17325
(717) 334-8632

Joel Guldin
Rd. 1 Synder Rd.
Green Lane 18054
(215) 234-4932

Sealand's Metal Detectors
422 Sells Lane
Greensburg 15601
(512) 834-3429

Miller's Treasures &
Coin Detectors
Rd. #1 Pettis Rd.
Meadville 16335
(814) 335-4936

James Elder
Rd. #1 Box 253-A
Mifflintown 17059

Fox Metal Detectors
313 Bridge St.
New Cumberland 17070
(717) 232-4629

Cavalier Metal Detector Sales
333 Carmell Dr.
Pittsburgh 15241
(412) 941-5691

Richard Haviland
3228 Brownsville Rd.
Pittsburgh 15227
(412) 882-8073

Prospect Metal Detectors
737 Second Ave.
Prospect Park 19076
(215) 534-1685

Mid State Sales
3410 W. College Ave.
State College 16801
(814) 238-5666

Rhode Island

House of Bargains
345 Warwick Ave.
Warwick 02888
(401) 781-8580

1Bob's Service
RFD 1 Box 135
Benefit St.
Greene 02827
(401) 397-3143

Tennessee

Hickory Valley Electric Co.
6916 Lee Hwy.
Chattanooga 37421
(615) 892-0525

C/S Detector Electronics
354 Miffin Rd. Rt 6
Jackson 38301
(901) 424-6319

The Collector's Shop
100 Oak Shopping Ctr.
Nashville 37204
(615) 383-5996

Texas

William B. Doss
3816 N. 11th
Abilene 79603
(915) 673-5102

Arlington Electronic Center
915 E. Park Row
Arlington 76010
(817) 261-9441

Arlington Treasure Chest
1735 S. Cooper
Arlington 76010
(817) 261-5272

Bill's Trailer Sales
1800 New Jersey
Baytown 77520
(713) 422-4322

M & M THer's Supply
Box 83
Bowie 76230
(817) 872-3439

Treasure Hound
Detector Sales
400 Mitchell
Bryan 77801

(713) 823-6423

Hide-N-Seek
101 Hwy. 288 S.
Clute 77531
265-6562

National Treasure Hunter's League
11602 Garland Rd.
Dallas 75218

El Paso Equipment Rental
6055 Alameda
El Paso 79905
(915) 778-3351

Joe Krajca & Son
Rt. 3
Ennis 75119
(214) 875-6358

Don Gray Enterprises
1561 W. Berry
Fort Worth 76110
(817) 921-2531

Rex Grove Auto Supply
4527 E. Belknap
Fort Worth 76117
(817) 838-3066

Jen-Mar Laundry & Cleaning
2805 Weiler
Fort Worth 76119
(817) 451-9005

Research & Recovery
2803 Old Spanish Trail
Houston 77054
(713) 747-4647

J. W. & A. Assoc. Inc.
3229 W. Pioneer
Irving 75061
(214) 252-7661

Noel & Betty Murchison
201 George
Lubbock 79416

Mesquite Craft & Hobby Mart
508 S. Galloway
Mesquite 75149
(214) 285-2930

Mission Drug Store
1030 Conway
Mission 78572
(512) 585-1532

The Treasure Bug Shop
Rt. 2 Box 511
Orange 77630
(713) 745-1190

Owen's Detector Sales
5814 Kepler Dr.
San Antonio 78228

San Antonio Mission Metal Detector Co.
2302 W. Hermosa Dr.
San Antonio 78201
(512) 732-5522

Utah

John Bylund Plb. & Htg.
1255 E. Center
Pleansant Grove 84062
(801) 785-3540

Bryant T. Cash
2457 W. 4975 S.
Roy 84067
(801) 825-7858

Gallenson's
220 S. State St.
Salt Lake City 84111

(901) 328-2016

Vermont

Leisure Lines
Woodstock Ave.
Rutland 05701
(802) 775-0854

Virginia

Suburban Detector Sales
3169 Spring St.
Fairfax 22030
(703) 273-2542

Chancellorsville Outpost
Rt. 1 Box 170
Fredericksburg 22401

Essential Instruments
10453 Medina Rd.
Richmond 23235
(804) 272-5558

Tabb Trailer Sales
1459 Hwy. 17
Tabb 23602
(804) 595-5720

H & S Detector Center
2214 Thoroughgood Rd.
Virginia Beach 23455
(804) 464-6072

Opequon Trader
807 Berryville
Winchester 22601
(703) 667-8257

Washington

The Treasure Chest
Rt. 1 Box 39-F
Granite Falls 98252
(206) 334-5263

The Coin Cradle
320 W. Kennewick Ave.
Kennewick 99336
(509) 586-4982

Doug's Hobby & Prospecting Shop
6908 214th S.W.
Lynnwood 98036

Pearl Electronics
1300 First Ave.
Seattle 98101
(206) 622-6200

Bob Carstensen
17 Whispering Firs
Silverdale 98386
(502) 692-6882

Bowen's Hideout
S. 1823 Mt. Vernon
Spokane 99203
(509) 534-4004

B & I Coin Shop
8012 S. Tacoma Way
Tacoma 98499
(206) 584-6241

Snake River Prospector's Supply
821 St. John St.
Walla Walla 99362
(509) 525-8304

West Virginia

Murdock's Hobby Shop
121 N. Fourth Ave.
Paden City 26159
(304) 337-2711

Wisconsin

Don's Treasure Hunting Supply
N. 88 W. 16747
Menomonee Falls 53051
(414) 251-5350

Nyland's, Inc.
1607 Main St.
Marinette 54143
(715) 735-5110

Keith Rasmussen
1536 Michigan Blvd.
Racine 53402
(414) 637-6658

Jetzer's Locksmith Service
1822 N. 12th St.
Sheboygan 53081
(414) 457-9231

Wyoming

Mike's Sporting Goods
Box 1505
Casper 82601
(307) 234-2294

Appendix 3
Selected Bibliography

PROSPECTING, MINERALS AND GEMS

The Gold Diver's Handbook, Matt Thornton
Handbook for Prospectors Richard M. Pearl
Diving and Digging for Gold, Mary Hill
Gold Finding Secrets, Edwin P. Morgan
Let's Go Prospecting, Edward Arthur
Prospecting and Operating Small Gold Placers, William F. Boericke
Gold: ABCs of Panning, E.S. LeGaye
Prospecting for Gold, W.F. Heinz
Gold Diggers' Atlas, Robert Neil Johnson
Prospecting Hints, United Prospectors, Inc.
Prospecting for Gemstones and Minerals, John Sinkankas
Prospecting for Lode Gold, Gregory Stone
Where to Find Gold in the Mother Lode, James Klein
Geology, Frank H. T. Rhodes
Landforms, Adams and Wyckoff
Rocks and Minerals, Herbert S. Zim and Paul R. Shaffer
A Field Guide to Rocks and Minerals, Frederick H. Pough
How to Know the Minerals and Rocks, Richard M. Pearl
Mineral Recognition, Iris Vanders and Paul F. Kerr
Dana's Manual of Mineralogy, Cornelius Hurlbut
Rocks and Minerals of California, Vinson Brown, et al.
Exploring and Mining for Gems and Gold in the West, Fred Rynerson

Gold Panning with Prospector John, Prospector John
Where to Find Gold in Southern California, James Klein
Everything You Wanted to Know About Gold, Russel Burkett
The Gold Miner's Pocket Companion, Bob Bigando, Jr.
Gold Locations of the United States, Jack Black
Minerals of the World, Charles Sorrell
How to Build and Operate the Sluice Box, T.R. Glenn
Where and How to Find Gold, Gems or Junk, J. Allen Hitchens
Handbook of Jade, G. Hemrich
Gem Minerals of Idaho, John A. Beckwith
Exploring Minerals and Crystals, Robert I. Gait
Hunting Diamonds in California, Mary Hill
Desert Gem Trails, Mary Frances Strong
Gemstones of North America, John Sinkankas
Gem Trails in California, A.L. Abbott
Gold, the 79th Element, S.R. Wendt
Gold Panner's Manual, Garnet Basque
Where to Find Gold in the Desert, James Klein

LAPIDARY

Tumbling Techniques, G.L. Daniel
Gem Cutting–A Lapidary's Manual, John Sinkankas
Jewelry Making by the Lost Wax Process, Greta Pack
Precious Stones, Max Bauer
Specialized Gem Cutting, Jack R. Cox

OLD MINES, GHOST TOWNS, AND MINING CAMPS

Ghost Towns of the West, Lambert Florin
Famous Lost Mines of the Old West, John Latham
Forgotten Mines and Treasures, Wayne Winters
Ghost Town Trails, Lambert Florin
Western Ghost Town Shadows, Lambert Florin
Ghost Town Treasures, Lambert Florin
Ghost Town El Dorado, Lambert Florin
Western Ghost Towns, Lambert Florin
A Guide to Western Ghost Towns, Lambert Florin
Colorado Ghost Towns–Past and Present, Robert L. Brown
Ghost Towns of the Colorado Rockies, Robert L. Brown
Jeep Trails to Colorado Ghost Towns, Robert L. Brown

Southwestern Ghost Town Atlas, Robert Neil Johnson
Ghost Towns and Mining Camps of California, Remi Nadeau
Ghost Towns of the Northwest, Norman D. Weis
California Nevada Ghost Town Atlas, Robert Neil Johnson
Nevada Ghost Towns and Mining Camps, Stanley W. Pahner
Nevada Ghost Towns, Raymond C. Browne
Mines of Julian, Helen Ellsberg
Mines of the San Gabriels, John W. Robinson
Mines of the High Desert, Ronald Dean Miller
Mines of Death Valley, L. Burr Belden
Mines of Eastern Sierra, Mary DeDecker
Mines of the Mojave, Ron and Peggy Miller
Ghost Towns of Arizona, James & Barbara Sherman
Exploring the Ghost Town Desert, Roberta Starry

TREASURE HUNTING AND METAL DETECTING

Treasure Hunter's Yearbook, A.T. Evans
Getting Started in Treasure Hunting, Alan Smith
The Treasure Hunter's Manual, Karl Von Meuller
The Modern Treasure Finder's Manual, George Sullivan
Electronic Detector Handbook, E.S. LeGaye
Treasure Hunting with the Metal Detector, Marvin and Helen Davis
Coinshooting–How and Where to Do It, H. Glenn Carson
THing: A Modern Search for Adventure, H. Glenn Carson
Backyard Treasure Hunting, Lucie Lowery
An Eastern THing, Ralph Edmonds
A Guide to Treasure in Texas, Thomas Penfield
A Guide to Treasure In Arkansas, Louisiana and Mississippi, Thomas Penfield
A Guide To Treasure in Arizona, Thomas Penfield
Canadian Treasure Trove, Garnet Basque
Campfires Along the Treasure Trail, Wayne Winters
Guidebook to Lost Western Treasure, Robert Fergueson
Lost Desert Gold, Ralph Caine
The Gold Hex, Ken Marquiss
The Sterling Legend, Estee Conaster
Pegleg–To Date and Beyond, John Southworth

HISTORY

Here Rolled the Covered Wagons, Albert and Jane Salisbury
Two Captains West, Albert and Jane Salisbury
Old Forts of the Southwest, Herbert M. Hart
Where the Washingtonians Lived, Lucile McDonald
Early Oregon, R. N. Preston
Western Wagon Wheels, Lambert Florin
Historic Western Churches, Lambert Florin
Washington, Bob and Ira Spring and Ron Fish
Exploring Historic California, Jack Adler
Western Mining, Otis E. Young, Jr.
Mining Camp Days, Emil W. Billeb
Comstock Mining and Miners, Eliot Lord
An Empire of Silver, Robert L. Brown
Gold Rushes and Mining Camps of the Early American West, Vardis Fisher and Opal L. Holmes
The Nevada Desert, Sessions S. Wheeler
Right-of-Way–A Guide to Abandoned Railroads in the United States, Waldo Nielsen
Gold Cities, Jim Morley and Doris Foley
Gold Mines of California, Jack R. Wagner
Exploring California Folklore, Russ Leadabrand
Travellers' Guide to the Comstock Lode, Arthur Lassagne

TRAVEL

Exploring California Byways, No. 2, Russ Leadabrand
Exploring California Byways, No.3, Russ Leadabrand
Exploring California Byways, No.4, Russ Leadabrand
Exploring California Byways, No. 5, Russ Leadabrand
Exploring California Byways, No.6, Russ Leadabrand
Exploring California Byways, No. 7, Russ Leadabrand
Exploring Small Towns, No. 1: Southern California, David Yeadon
Guidebook to Missions of California, Marjorie Camphouse
A Guidebook to the San Gabriel Mountains of California, Russ Leadabrand
A Guidebook to the Southern Sierra Nevada, Russ Leadabrand
A Guidebook to the San Jacinto Mountains, Russ Leadabrand
A Guidebook to the Mountains of San Diego and Orange Counties, Russ Leadabrand

A Guidebook to the San Bernardino Mountains of California, Russ Leadabrand
Where to take Your Children In Southern California, Davis Dutton and Tedi Pilgreen
A Guidebook to the Northern California Coast, Vol. 1, Mike Hayden
A Guidebook to the Northern California Coast, Vol. II, Mike Hayden
A Guidebook to the Mojave Desert of California, Russ Leadabrand
Guidebook to the Feather River Country, Jim Martin
Exploring Big Sur, Monterey and Carmel, Maxine Knox and Mary Rodriguez
Exploring Death Valley, Ruth Kirk
Exploring the Mother Lode Country, Richard Dillon
Gold Rush Country, Editors of Sunset Books and Magazine
California State Parks, John Robinson and Alferd Calais
Back Roads of California, Earl Thollander
Close Ups of High Sierras, Norman Clyde
Death Valley Jeep Trails, Roger Mitchell
Eastern Sierra Jeep Trails, Roger Mitchell
Inyo-Mono Jeep Trails, Roger Mitchell
Western Nevada Jeep Trails, Roger Mitchell
Unbeaten Paths, Ron Fish
Grand Canyon Treks, Dr. Harvey Butchart
Byroads of Baja, Walt Wheelock
Camping and Climbing in Baja, John W. Robinson
Baja California, Choral Pepper
Beaches of Sonora, Walt Wheelock

BOTTLES, RELICS, AND ARTIFACTS

Old Time Bottles, Lynn Blumenstein
Redigging the West, Lynn Blumenstein
Bottle Rush USA, Lynn Blumenstein
Ghost Town Bottle Pricing Guide, Wes and Ruby Bressie
Pocket Field Guide for the Bottle Digger, Marvin & Helen Davis
Milk Bottle Manual, Gordon A. Taylor
Ghost Town Relics, Wes and Ruby Bressie
Treasure Hunter's Relic Identification, Lynn Blumenstein
Truly American, Lynn Blumenstein
The Golden Guide to American Antiques, Ann Kilborn Cole
Treasury of Frontier Relics, Les Beitz

Relics of the Redman, Marvin and Helen Davis
Relics of the Whiteman, Marvin D. Davis

COINS AND CURRENCY

Guide Book of U.S. Coins, R.S. Yeoman
Modern United States Currency, Neil Shafer
Current Coins of the World, R.S. Yeoman
Modern World of Coins, R.S. Yeoman
Buying and Selling U.S. Coins, Ken Bressett

FOSSILS

An Introduction to Paleobotany, C.A. Arnold
Evolution of the Vertebrates, E.H. Colbert
Fossils, W.H. Matthews
Introduction to Historical Geology, R.C. Moore

Index